AF556192

The Chemistry and Biology of Leaf

The Chemistry and Biology of Leaf

Dr. Satish Kumar Sinha

RANDOM PUBLICATIONS

NEW DELHI - 110 002 (INDIA)

The Chemistry and Biology of Leaf

ISBN 978-93-51114-52-9

Published in 2014 in India by

RANDOM PUBLICATIONS

4376-A/4B, Gali Murari Lal, Ansari Road
New Delhi-110 002
Phone: +9111-43580356, 23289044
E-mail: randomexports@gmail.com; sales@randompublications.com; info@randompublications.com

Type Setting by: Friends Media, Delhi-110089
Printed at Thomson Press (India) Ltd

Preface

A leaf is an organ of a vascular plant, as defined in botanical terms, and in particular in plant morphology. Foliage is a mass noun that refers to leaves as a feature of plants. Typically a leaf is a thin, flattened organ borne above ground and specialized for photosynthesis, but many types of leaves are adapted in ways almost unrecognisable in those terms: some are not flat, some are not above ground, and some are without major photosynthetic function. Conversely, many structures of non-vascular plants, or even of some lichens, which are not plants at all, do look and function much like leaves. Furthermore, several structures found in vascular plants look like leaves but are not totally homologous with leaves; they differ from typical leaves in their structures and origins. Examples include phyllodes, cladodes, and phylloclades.

Typically leaves are flat and thin, thereby maximising the surface area directly exposed to light and promoting photosynthetic function. Externally they commonly are arranged on the plant in such ways as to expose their surfaces to light as efficiently as possible without shading each other, but there are many exceptions and complications; for instance plants adapted to windy conditions may have pendent leaves, such as in many willows and Eucalyptus. Likewise, the internal organisation of most kinds of leaves has evolved to maximise exposure of the photosynthetic organelles, the chloroplasts, to light and to increase the absorption of carbon dioxide. Most leaves have stomata, which open or narrow to regulate the exchange of carbon dioxide, oxygen, and water vapour with the atmosphere. In contrast however, some leaf forms are adapted to modulate the amount of light they absorb to avoid or mitigate excessive heat, ultraviolet damage, or desiccation, or to sacrifice light-absorption efficiency in favour of protection from herbivorous enemies. Among these forms the leaves of many xerophytes are conspicuous. For such plants their major constraint is not light flux or intensity, but heat, cold, drought, wind, herbivory, and various other hazards. Typical examples among such strategies are so-called window plants such as Fenestraria species, some Haworthia species such as Haworthia tesselata and Haworthia truncata and Bulbine mesembryanthemoides. The shape and structure of leaves vary considerably from species to species of plant, depending largely on their adaptation to climate and available light, but also to other factors such as grazing animals, available nutrients, and ecological competition from other plants. Considerable changes in leaf type occur within species too, for example as a plant matures; as a case in point Eucalyptus species commonly have isobilateral, pendent leaves when mature and dominating their neighbours; however, such trees tend to have erect or horizontal dorsiventral leaves as seedlings, when their growth is limited by the available light. Other factors include the need to balance water loss at high temperature and low humidity against the need to absorb atmospheric carbon dioxide. In most plants leaves also are the primary organs responsible for transpiration and guttation. Leaves can also store food and water, and are modified accordingly to meet these functions, for example in the leaves of succulent plants and in bulb scales. The concentration of photosynthetic structures in leaves requires that they be richer in protein, minerals, and sugars, than say, woody stem tissues. Accordingly leaves are prominent in the diet of many animals. This is true for humans, for whom leaf vegetables commonly are food staples. Deciduous plants in frigid or cold temperate regions typically shed their leaves in autumn, whereas in areas with a severe dry season, some plants may shed their leaves until the dry season ends. In either case the shed leaves may be expected to contribute their retained nutrients to the soil where they fall. In contrast, many other non-seasonal plants, such as palms and conifers, retain their leaves for long periods; Welwitschia retains its two main leaves throughout a lifetime that may exceed a thousand years.

The present book will be a standard reference for the formal analysis of photosynthetic metabolism in vivo by advanced students and researchers.

I thank all members of my team who have helped in the preparation of the book. My special thanks go to "Random Publications" who have published the book.

— Dr. Satish Kumar Sinha

Contents

1. Starting the Story 1

- General Nature of Leaves 1
- Seasonal Leaf Loss 8
- Divisions of the Blade 10
- Interactions with Other Organisms 19
- Basic Types 22
- Organisms Confused with Ferns 32
- Frond 32
- Lycopodiophyta 40
- Emergence 44
- Succulent Plant 49
- Cataphyll 51
- Bulb-Scales 55

2. Seeking Illumination 58

- Autumn Leaf Colour 58
- Pigments that Contribute to Other Colours 60

- Cell Walls 62
- Allelopathy 63
- Deciduous Woody Plants 67

3. Diffusing Gases 83

- Gas Exchange 83
- Leaves 85
- Density of Stomata 87
- Roots and Stems 88

4. Leaking Water 92

- Stoma 92
- Function 92
- Stomata and Climate Change 96
- Regulation 100
- Antitranspirant 103
- Soil Plant Atmosphere Continuum 103
- Energy Balance 107

5. Raising Water 121

- Role of Endodermis 121
- Importance 122
- How does Water Move through Plants to Get to the Top of Tall Trees? 125

- Mechanism Driving Water Movement in Plants 133
- Transpiration Stream 138

6. Interfacing with Air **140**

- Photosynthesis 140
- Chloroplast DNA 154
- DNA Replication 155
- Thylakoid 177
- Thylakoid Function 183
- Thylakoid Membranes in Cyanobacteria 185
- Light-Dependent Reactions 186
- Photodissociation 195
- Oxygen Evolution 200
- Carbon Fixation 202
- Photosynthetic Efficiency 208
- Efficiencies of Various Biofuel Crops 211
- Evolution of Photosynthesis 212
- Photorespiration 219

7. Keeping Cool **228**

- Keep It Cool: What Desert Plants can Teach Us about Climate Change 230
- What Strategies Allow Desert Plants to Survive Harsh Conditions? 231
- How to Help Your Crops Endure Hot Weather 232

8. **Cleaning Surfaces** **234**

- Lotus Effect 234
- Functional Principle 234
- How Nature Cleans 237
- Rice Leaves and Butterfly Wings Provide Insight into Nature's Best Self-Cleaning Surfaces 239
- Lotus Leaves, Hydrophobic and Omniphobic Surfaces 242

9. **Staying Unfrozen** **245**

- Do Insects Hibernate? 245
- Vernalization 246
- Vernalization in Arabidopsis Thaliana 248
- Antifreeze Protein 248
- Diversity 250
- Mechanisms of Action 252
- Dehydrin 254

10. **Staying Stiff and High** **256**

- Plants Exhibit a Wide Range of Mechanical Properties, Engineers Find 260
- Which Plants will Survive Droughts, Climate Change? 262

11. **Surviving a Storm** **267**

- Understanding Lightning and Associated Tree Damage 267

- Lightning Protection Systems for Trees 268
- Trees and Our Climate 270

12. Making and Maintaining **272**

- Benefits of Trees 272
- Environmental Benefits 273
- Trees Require an Investment 275
- Tree Function 279
- Bare-Root Stock 284
- Medium Zone 290
- Mature Tree Care 295
- Pruning Techniques 307
- Helping Trees Recover from Stress 309

Bibliography 317

Index 319

1

Starting the Story

A leaf is an organ of a vascular plant, as defined in botanical terms, and in particular in plant morphology. Foliage is a mass noun that refers to leaves as a feature of plants.

Typically a leaf is a thin, flattened organ borne above ground and specialised for photosynthesis, but many types of leaves are adapted in ways almost unrecognisable in those terms: some are not flat (for example many succulent leaves and conifers), some are not above ground (such as bulb scales), and some are without major photosynthetic function (consider for example cataphylls, spines, and cotyledons).

Conversely, many structures of non-vascular plants, or even of some lichens, which are not plants at all (in the sense of being members of the kingdom Plantae), do look and function much like leaves. Furthermore, several structures found in vascular plants look like leaves but are not totally homologous with leaves; they differ from typical leaves in their structures and origins. Examples include phyllodes, cladodes, and phylloclades.

According to Agnes Arber's partial-shoot theory of the leaf, leaves are partial shoots. Compound leaves are closer to shoots than simple leaves. Developmental studies have shown that compound leaves, like shoots, may branch in three dimensions. On the basis of molecular genetics, Eckardt and Baum (2010) concluded that "is is now generally accepted that compound leaves express both leaf and shoot properties."

General Nature of Leaves

Typically leaves are flat and thin, thereby maximising the surface area directly exposed to light and promoting photosynthetic function. Externally they commonly are arranged on the plant in such ways as to expose their surfaces to light as efficiently as possible without shading each other, but there are many exceptions and complications; for instance plants adapted to windy conditions may have pendent leaves, such as in many willows and *Eucalyptus*.

Likewise, the internal organisation of most kinds of leaves has evolved to maximise exposure of the photosynthetic organelles, the chloroplasts, to light and to increase the absorption of carbon dioxide. Most leaves have stomata, which open or narrow to regulate the exchange of carbon dioxide, oxygen, and water vapour with the atmosphere.

In contrast however, some leaf forms are adapted to modulate the amount of light they absorb to avoid or mitigate excessive heat, ultraviolet damage, or desiccation, or to sacrifice light-absorption efficiency in favour of protection from herbivorous enemies. Among these forms the leaves of many xerophytes are conspicuous.

For such plants their major constraint is not light flux or intensity, but heat, cold, drought, wind, herbivory, and various other hazards. Typical examples among such strategies are so-called window plants such as Fenestraria species, some Haworthia species such as Haworthia tesselata and Haworthia truncata and Bulbine mesembryanthemoides.

The shape and structure of leaves vary considerably from species to species of plant, depending largely on their adaptation to climate and available light, but also to other factors such as grazing animals, available nutrients, and ecological competition from other plants. Considerable changes in leaf type occur within species too, for example as a plant matures; as a case in point *Eucalyptus* species commonly have isobilateral, pendent leaves when mature and dominating their neighbours; however, such trees tend to have erect or horizontal dorsiventral leaves as seedlings, when their growth is limited by the available light.

Other factors include the need to balance water loss at high temperature and low humidity against the need to absorb atmospheric carbon dioxide. In most plants leaves also are the primary organs responsible for transpiration and guttation (beads of fluid forming at leaf margins).

Leaves can also store food and water, and are modified accordingly to meet these functions, for example in the leaves of succulent plants and in bulb scales. The concentration of photosynthetic structures in leaves requires that they be richer in protein, minerals, and sugars, than say, woody stem tissues. Accordingly leaves are prominent in the diet of many animals. This is true for humans, for whom leaf vegetables commonly are food staples.

Figure: *A leaf shed in autumn.*

Correspondingly, leaves represent heavy investment on the part of the plants bearing them, and their retention or disposition are the subject of elaborate strategies for dealing with pest pressures, seasonal conditions, and protective measures such as the growth of thorns and the production of phytoliths, lignins, tannins and poisons.

Deciduous plants in frigid or cold temperate regions typically shed their leaves in autumn, whereas in areas with a severe dry season, some plants may shed their leaves until the dry season ends. In either case the shed leaves may be expected to contribute their retained nutrients to the soil where they fall.

In contrast, many other non-seasonal plants, such as palms and conifers, retain their leaves for long periods; *Welwitschia* retains its two main leaves throughout a lifetime that may exceed a thousand years.

Not all plants have true leaves. Bryophytes (e.g., mosses and liverworts) are non-vascular plants, and, although they produce flattened, leaf-like structures that are rich in chlorophyll, these organs differ morphologically from the leaves of vascular plants; For one thing, they lack vascular tissue. Vascularised leaves first evolved following the Devonian period, when carbon dioxide concentration in the atmosphere dropped significantly. This occurred independently in two separate lineages of vascular plants: the microphylls of lycophytes and the euphylls ("true leaves") of ferns, gymnosperms, and angiosperms. Euphylls are also referred to as macrophylls or megaphylls ("large leaves").

Large-Scale Features (Leaf Morphology)

A structurally complete leaf of an angiosperm consists of a petiole (leaf stalk), a lamina (leaf blade), and stipules (small structures located to either side of the base of the petiole). Not every species produces leaves with all of these structural components. In certain species, paired stipules are not obvious or are absent altogether. A petiole may be absent, or the blade may not be laminar (flattened). The tremendous variety shown in leaf structure (anatomy) from species to species is presented in detail below under morphology. The petiole mechanically links the leaf to the plant and provides the route for transfer of water and sugars to and from the leaf. The lamina is typically the location of the majority of photosynthesis. The upper (adaxial) angle between a leaf and a stem is known as the axil of the leaf. It is often the location of a bud. Structures located there are called "axillary".

Anatomy

Medium Scale Features: Leaves are normally extensively vascularised and are typically covered by a dense network of xylem, which supply water for photosynthesis, and phloem, which remove the sugars produced by photosynthesis. Many leaves are covered in trichomes (small hairs) which have a diverse range of structures and functions.

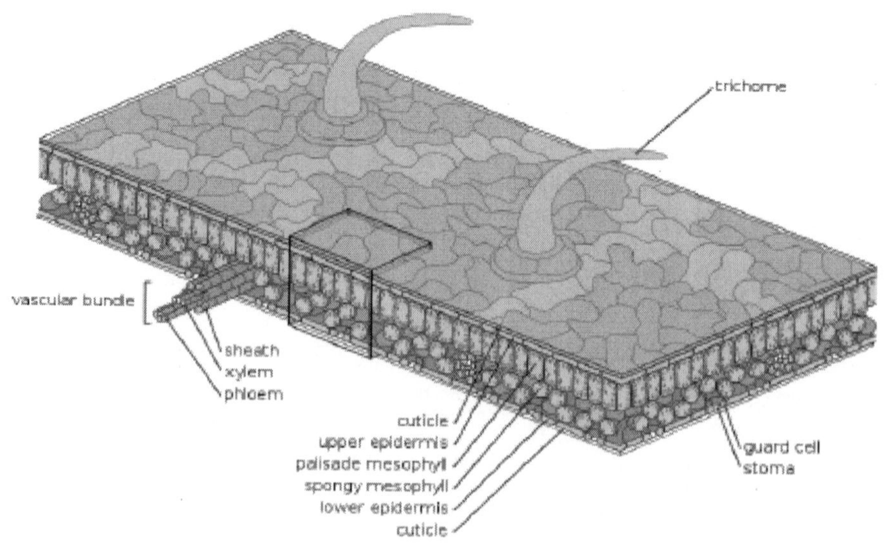

Small-Scale Features

A leaf is a plant organ and is a collection of tissues in a regular organisation. The major tissue systems present are

1. The epidermis, which covers the upper and lower surfaces
2. The mesophyll tissue inside the leaf, which is rich in chloroplasts (also called chlorenchyma)
3. The arrangement of veins (the vascular tissue)

These three tissue systems typically form a regular organisation at the cellular scale.

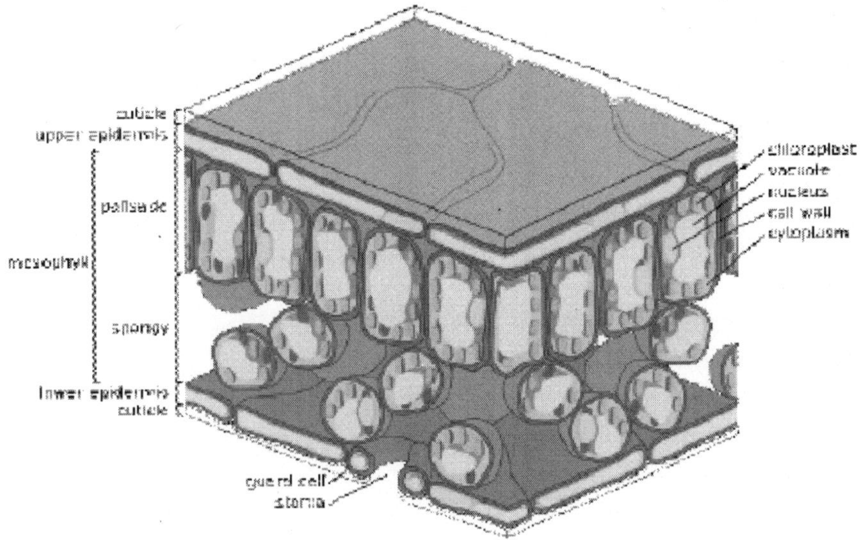

Major Leaf Tissues

Epidermis: The epidermis is the waxy outer layer of cells covering the leaf. It forms the boundary separating the plant's inner cells from the external world. The epidermis serves several functions: protection against water loss by way of transpiration, regulation of gas exchange, secretion of metabolic compounds, and (in some species) absorption of water. Most leaves show dorsoventral anatomy: The upper (adaxial) and lower (abaxial) surfaces have somewhat different construction and may serve different functions.

The epidermis is usually transparent (epidermal cells lack chloroplasts) and coated on the outer side with a waxy cuticle that prevents water loss. The cuticle is in some cases thinner on the lower epidermis than on the upper epidermis, and is generally thicker on leaves from dry climates as compared with those from wet climates.

The epidermis tissue includes several differentiated cell types: epidermal cells, epidermal hair cells (trichomes) cells in the stomate complex; guard cells and subsidiary cells. The epidermal cells are the most numerous, largest, and least specialised and form the majority of the epidermis. These are typically more elongated in the leaves of monocots than in those of dicots.

The epidermis is covered with pores called *stomata*, part of a stoma complex consisting of a pore surrounded on each side by chloroplast-containing guard cells, and two to four subsidiary cells that lack chloroplasts. Opening and closing of the stoma complex regulates the exchange of gases and water vapor between the outside air and the interior of the leaf and plays an important role in allowing photosynthesis without letting the leaf dry out. In a typical leaf, the stomata are more numerous over the abaxial (lower) epidermis than the adaxial (upper) epidermis and more numerous in plants from cooler climates.

Mesophyll

Most of the interior of the leaf between the upper and lower layers of epidermis is a *parenchyma* (ground tissue) or *chlorenchyma* tissue called the mesophyll (Greek for "middle leaf"). This assimilation tissue is the primary location of photosynthesis in the plant. The products of photosynthesis are called "assimilates".

In ferns and most flowering plants, the mesophyll is divided into two layers:

- An upper palisade layer of tightly packed, vertically elongated cells, one to two cells thick, directly beneath the adaxial epidermis. Its cells contain many more chloroplasts than the spongy layer. These long cylindrical cells are regularly arranged in one to five rows. Cylindrical cells, with the *chloroplasts* close to the walls of the cell, can take optimal advantage of light. The slight separation of the cells provides maximum absorption of carbon dioxide. This separation must be minimal to afford capillary action for water distribution. In order to adapt to their different environment (such as sun or shade), plants had to adapt this structure to obtain optimal result. Sun leaves have a multi-layered palisade layer, while shade leaves or older leaves closer to the soil are single-layered.
- Beneath the palisade layer is the spongy layer. The cells of the spongy layer are more rounded and not so tightly packed. There are large intercellular air spaces. These cells contain

fewer chloroplasts than those of the palisade layer. The pores or *stomata* of the epidermis open into substomatal chambers, which are connected to the air spaces between the spongy layer cells.

These two distinct layers of the mesophyll are absent in many aquatic and marsh plants. Even an epidermis and a mesophyll may be lacking. Instead, for their gaseous exchanges they use a homogeneous aerenchyma (thin-walled cells separated by large gas-filled spaces). Their stomata are situated at the upper surface.

Leaves are normally green, due to chlorophyll in plastids in the chlorenchyma cells. Plants that lack chlorophyll cannot photosynthesize optimally. Photosynthesis can still be performed utilising other pigments such as carotenes and xanthophylls.

Veins

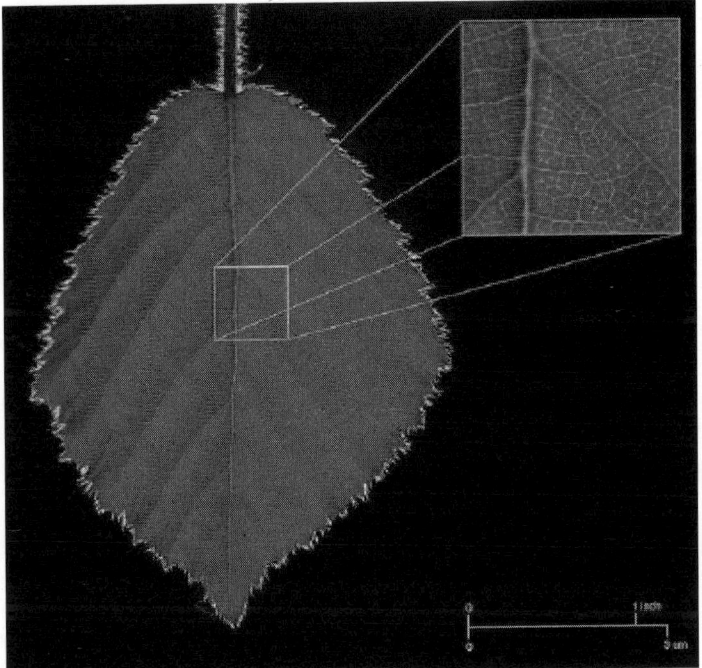

Figure: *The veins of a bramble leaf*

The veins are the vascular tissue of the leaf and are located in the spongy layer of the mesophyll. The pattern of the veins is called venation, and is typically characterized by hierarchical structures with abundant closed loops. They were once thought to be typical examples of pattern formation through ramification, but they may instead exemplify a pattern formed in a stress tensor field. A vein is made up of a vascular bundle. At the core of each bundle are clusters of two distinct types of ducts (tubes):

- Xylem: ducts that bring water and minerals from the roots into the leaf.
- Phloem: ducts that usually move sap, with dissolved sucrose, produced by photosynthesis in the leaf, out of the leaf.
- A sheath of ground tissue made of lignin surrounding the ducts. This sheath has a mechanical role in strengthening the rigidity of the leaf.

The xylem typically lies on the adaxial side of the vascular bundle and the phloem typically lies on the abaxial side. Both are embedded in a dense parenchyma tissue, called the pith or sheath, which usually includes some structural collenchyma tissue.

Seasonal Leaf Loss

Leaves in temperate, boreal, and seasonally dry zones may be seasonally deciduous (falling off or dying for the inclement season). This mechanism to shed leaves is called abscission. After the leaf is shed, a leaf scar develops on the twig. In cold autumns, they sometimes change colour, and turn yellow, bright-orange, or red, as various accessory pigments (carotenoids and xanthophylls) are revealed when the tree responds to cold and reduced sunlight by curtailing chlorophyll production. Red anthocyanin pigments are now thought to be produced in the leaf as it dies, possibly to mask the yellow hue left when the chlorophyll is lost—yellow leaves appear to attract herbivores such as aphids. Optical masking of chlorophyll by anthocyanins reduces risk of photo-oxidative damage to leaf cells as they senesce, which otherwise may lower the efficiency of nutrient retrieval from senescing autumn leaves.

Morphology

External leaf characteristics (such as shape, margin, hairs, etc.) are important for identifying plant species, and botanists have developed a rich terminology for describing leaf characteristics. These structures are a part of what makes leaves determinant; they grow and achieve a specific pattern and shape, then stop. Other plant parts like stems or roots are non-determinant, and will usually continue to grow as long as they have the resources to do so.

Classification of leaves can occur through many different designative schema, and the type of leaf is usually characteristic of a species, although some species produce more than one type of leaf. The longest type of leaf is a leaf from a palm, measuring at nine feet long. The terminology associated with the description of leaf morphology is presented, in illustrated form, at Wikibooks.

Arrangement on the Stem

Different terms are usually used to describe leaf placement (phyllotaxis):

Figure: *The leaves on this plant are arranged in pairs opposite one another, with successive pairs at right angles to each other ("decussate") along the red stem. Note the developing buds in the axils of these leaves.*

- Alternate – leaf attachments are singular at nodes, and leaves alternate direction, to a greater or lesser degree, along the stem.
- Basal – arising from the base of the stem.
- Cauline – arising from the aerial stem.
- Opposite – Two structures, one on each opposite side of the stem, typically leaves, branches, or flower parts. Leaf attachments are paired at each node and decussate if, as typical, each successive pair is rotated 90° progressing along the stem.
- Whorled – three or more leaves attach at each point or node on the stem. As with opposite leaves, successive whorls may or may not be decussate, rotated by half the angle between the leaves in the whorl (i.e., successive whorls of three rotated 60°, whorls of four rotated 45°, etc.). Opposite leaves may appear whorled near the tip of the stem.

- Rosulate – leaves form a rosette
- Rows – The term "distichous" literally means "two rows". Leaves in this arrangement may be alternate or opposite in their attachment. The term "2-ranked" is equivalent. The terms tristichous and tetrastichous are sometimes encountered. For example, the "leaves" (actually microphylls) of most species of *Selaginella* are tetrastichous, but not decussate.

As a *stem* grows, leaves tend to appear arranged around the stem in a way that optimizes yield of light. In essence, leaves form a helix pattern centred around the stem, either clockwise or counterclockwise, with (depending upon the species) the same angle of divergence. There is a regularity in these angles and they follow the numbers in a Fibonacci sequence: 1/2, 2/3, 3/5, 5/8, 8/13, 13/21, 21/34, 34/55, 55/89. This series tends to a limit close to 360° × 34/89 = 137.52° or 137° 302 , an angle known in mathematics as the golden angle. In the series, the numerator indicates the number of complete turns or "gyres" until a leaf arrives at the initial position and the denominator indicates the number of leaves in the arrangement. This can be demonstrated by the following:

- alternate leaves have an angle of 180° (or 1/2)
- 120° (or 1/3) : three leaves in one circle
- 144° (or 2/5) : five leaves in two gyres
- 135° (or 3/8) : eight leaves in three gyres.

Divisions of the Blade

Figure: *A leaf with laminar structure and pinnate venation*

Two basic forms of leaves can be described considering the way the blade (lamina) is divided. A simple leaf has an undivided blade. However, the leaf shape may be formed of lobes, but the gaps between lobes do not reach to the main vein.

A compound leaf has a fully subdivided blade, each leaflet of the blade separated along a main or secondary vein. Because each leaflet can appear to be a simple leaf, it is important to recognise where the petiole occurs to identify a compound leaf. Compound leaves are a characteristic of some families of higher plants, such as the Fabaceae. The middle vein of a compound leaf or a frond, when it is present, is called a rachis.

- *Palmately compound* leaves have the leaflets radiating from the end of the petiole, like fingers of the palm of a hand, e.g. *Cannabis* (hemp) and *Aesculus* (buckeyes).
- *Pinnately compound* leaves have the leaflets arranged along the main or mid-vein.
    - o odd pinnate: with a terminal leaflet, e.g. *Fraxinus* (ash).
    - o even pinnate: lacking a terminal leaflet, e.g. *Swietenia* (mahogany).
- *Bipinnately compound* leaves are twice divided: the leaflets are arranged along a secondary vein that is one of several branching of the rachis. Each leaflet is called a "pinnule". The pinnules on one secondary vein are called "pinna"; e.g. *Albizia* (silk tree).
- *trifoliate* (or *trifoliolate*): a pinnate leaf with just three leaflets, e.g. *Trifolium* (clover), *Laburnum* (laburnum).
- *pinnatifid*: pinnately dissected to the central vein, but with the leaflets not entirely separate, e.g. *Polypodium*, some *Sorbus* (whitebeams). In pinnately veined leaves the central vein in known as the *midrib*.

Characteristics of the Petiole

Petiolated leaves have a petiole (leaf stem). Sessile (Epetiolate) leaves do not: The blade attaches directly to the stem. Subpetiolate leaves are nearly petiolate, or have an extremely short petiole, and appear sessile. In clasping or decurrent leaves, the blade partially or wholly surrounds the stem, often giving the impression that the shoot grows through the leaf. When this is the case, the leaves are called "perfoliate", such as in *Claytonia perfoliata*. In peltate leaves, the petiole attaches to the blade inside from the blade margin.

In some *Acacia* species, such as the Koa Tree (*Acacia koa*), the petioles are expanded or broadened and function like leaf blades; these are called phyllodes. There may or may not be normal pinnate leaves at the tip of the phyllode.

A stipule, present on the leaves of many dicotyledons, is an appendage on each side at the base of the petiole resembling a small leaf. Stipules may be lasting and not be shed (a stipulate leaf, such as in roses and beans), or be shed as the leaf expands, leaving a stipule scar on the twig (an exstipulate leaf).

- The situation, arrangement, and structure of the stipules is called the "stipulation".
  - free
  - adnate : fused to the petiole base
  - ochreate : provided with ochrea, or sheath-formed stipules, e.g. rhubarb,
  - encircling the petiole base
  - interpetiolar : between the petioles of two opposite leaves.
  - intrapetiolar : between the petiole and the subtending stem

Venation

There are two subtypes of venation, namely, *craspedodromous*, where the major veins stretch up to the margin of the leaf, and *camptodromous*, when major veins extend close to the margin, but bend before they intersect with the margin.

- Feather-veined, reticulate (also called pinnate-netted, penniribbed, penninerved, or penniveined) – the veins arise pinnately from a single mid-vein and subdivide into veinlets. These, in turn, form a complicated network. This type of venation is typical for (but by no means limited to) dicotyledons.
- Three main veins branch at the base of the lamina and run essentially parallel subsequently, as in *Ceanothus*. A similar pattern (with 3-7 veins) is especially conspicuous in Melastomataceae.
- Palmate-netted, palmate-veined, fan-veined; several main veins diverge from near the leaf base where the petiole attaches, and radiate toward the edge of the leaf, e.g. most *Acer* (maples).
- Parallel-veined, parallel-ribbed, parallel-nerved, penniparallel – veins run parallel for the length of the leaf, from the base to the apex. Commissural veins (small veins) connect the major parallel veins. Typical for most monocotyledons, such as grasses.

Figure: *Palmate-veined leaf*

- Dichotomous – There are no dominant bundles, with the veins forking regularly by pairs; found in *Ginkgo* and some pteridophytes.

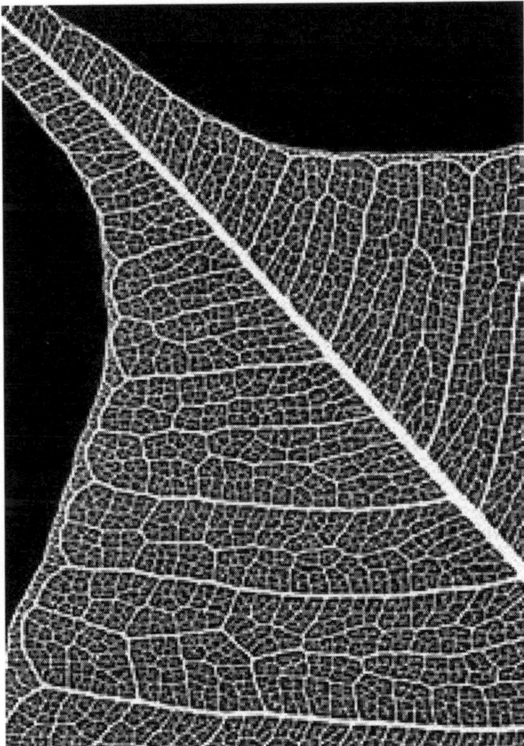

Figure: *Micrograph of a leaf-skeleton.*

Figure: *Dew on a leaf*

Note that, although it is the more complex pattern, branching veins appear to be plesiomorphic and in some form were present in ancient seed plants as long as 250 million years ago. A pseudo-reticulate venation that is actually a highly modified penniparallel one is an autapomorphy of some Melanthiaceae, which are monocots, e.g. *Paris quadrifolia* (True-lover's Knot).

Morphology Changes within a Single Plant

- Homoblasty – Characteristic in which a plant has small changes in leaf size, shape, and growth habit between juvenile and adult stages.
- Heteroblasty – Characteristic in which a plant has marked changes in leaf size, shape, and growth habit between juvenile and adult stages.

Edge (Margin)

- *ciliate:* fringed with hairs
- *crenate:* wavy-toothed; dentate with rounded teeth, such as *Fagus* (beech)
- *crenulate* finely or shallowly crenate
- *dentate:* toothed, such as *Castanea* (chestnut)
  - o *coarse-toothed:* with large teeth
  - o *glandular toothed:* with teeth that bear glands.

- *denticulate:* finely toothed
- *doubly toothed:* each tooth bearing smaller teeth, such as *Ulmus* (elm)
- *entire*: even; with a smooth margin; without toothing
- *linear*: parallel margins, elongated
- *lobate:* indented, with the indentations not reaching to the centre, such as many *Quercus* (oaks)
    - o *palmately lobed:* indented with the indentations reaching to the centre, such as *Humulus* (hop).
- *serrate:* saw-toothed with asymmetrical teeth pointing forward, such as *Urtica* (nettle)
- *serrulate:* finely serrate
- *sinuate:* with deep, wave-like indentations; coarsely crenate, such as many *Rumex* (docks)
- *spiny* or *pungent:* with stiff, sharp points, such as some *Ilex* (hollies) and *Cirsium* (thistles).

Tip

Figure: *Leaves showing various morphologies. Clockwise from upper left: tripartite lobation, elliptic with serrulate margin, peltate with palmate venation, acuminate odd-pinnate (centre), pinnatisect, lobed, elliptic with entire margin*

- *acuminate:* long-pointed, prolonged into a narrow, tapering point in a concave manner.
- *acute:* ending in a sharp, but not prolonged point
- *cuspidate:* with a sharp, elongated, rigid tip; tipped with a cusp.
- *emarginate:* indented, with a shallow notch at the tip.
- *mucronate:* abruptly tipped with a small short point, as a continuation of the midrib; tipped with a mucro.
- *mucronulate:* mucronate, but with a noticeably diminutive spine, a mucronule.
- *obcordate:* inversely heart-shaped, deeply notched at the top.
- *obtuse:* rounded or blunt
- *truncate:* ending abruptly with a flat end, that looks cut off.

Base

- *acuminate:* coming to a sharp, narrow, prolonged point.
- *acute:* coming to a sharp, but not prolonged point.
- *auriculate:* ear-shaped.
- *cordate:* heart-shaped with the notch towards the stalk.
- *cuneate:* wedge-shaped.
- *hastate:* shaped like an halberd and with the basal lobes pointing outward.
- *oblique:* slanting.
- *reniform:* kidney-shaped but rounder and broader than long.
- *rounded:* curving shape.
- *sagittate:* shaped like an arrowhead and with the acute basal lobes pointing downward.
- *truncate:* ending abruptly with a flat end, that looks cut off.

Surface

- *farinose:* bearing farina; mealy, covered with a waxy, whitish powder.
- *glabrous:* smooth, not hairy.
- *glaucous:* with a whitish bloom; covered with a very fine, bluish-white powder.
- *glutinous:* sticky, viscid.
- *papillate*, or *papillose:* bearing papillae (minute, nipple-shaped protuberances).

- *pubescent:* covered with erect hairs (especially soft and short ones).
- *punctate:* marked with dots; dotted with depressions or with translucent glands or coloured dots.
- *rugose:* deeply wrinkled; with veins clearly visible.
- *scurfy:* covered with tiny, broad scalelike particles.
- *tuberculate:* covered with tubercles; covered with warty prominences.
- *verrucose:* warted, with warty outgrowths.
- *viscid,* or *viscous:* covered with thick, sticky secretions.

The leaf surface is also host to a large variety of microorganisms; in this context it is referred to as the phyllosphere.

Hairiness

"Hairs" on plants are properly called trichomes. Leaves can show several degrees of hairiness. The meaning of several of the following terms can overlap.

- *arachnoid*, or *arachnose:* with many fine, entangled hairs giving a cobwebby appearance.
- *barbellate:* with finely barbed hairs (barbellae).
- *bearded:* with long, stiff hairs.
- *bristly:* with stiff hair-like prickles.
- *canescent:* hoary with dense grayish-white pubescence.
- *ciliate:* marginally fringed with short hairs (cilia).
- *ciliolate:* minutely ciliate.
- *floccose:* with flocks of soft, woolly hairs, which tend to rub off.
- *glabrescent:* losing hairs with age.
- *glabrous:* no hairs of any kind present.
- *glandular:* with a gland at the tip of the hair.
- *hirsute:* with rather rough or stiff hairs.
- *hispid:* with rigid, bristly hairs.
- *hispidulous:* minutely hispid.
- *hoary:* with a fine, close grayish-white pubescence.
- *lanate*, or *lanose:* with woolly hairs.
- *pilose:* with soft, clearly separated hairs.
- *puberulent*, or *puberulous:* with fine, minute hairs.
- *pubescent:* with soft, short and erect hairs.

- *scabrous*, or *scabrid:* rough to the touch.
- *sericeous:* silky appearance through fine, straight and appressed (lying close and flat) hairs.
- *silky:* with adpressed, soft and straight pubescence.
- *stellate*, or *stelliform:* with star-shaped hairs.
- *strigose:* with appressed, sharp, straight and stiff hairs.
- *tomentose:* densely pubescent with matted, soft white woolly hairs.
  - o *cano-tomentose:* between canescent and tomentose.
  - o *felted-tomentose:* woolly and matted with curly hairs.
- *tomentulose:* minutely or only slightly tomentose.
- *villous:* with long and soft hairs, usually curved.
- *woolly:* with long, soft and tortuous or matted hairs.

Timing

- *hysteranthous* developing after the flowers
- *synanthous* developing at the same time as the flowers

Adaptations

In the course of evolution, leaves have adapted to different environments in the following ways:

- A certain surface structure avoids moistening by rain and contamination.
- Sliced leaves reduce wind resistance.
- Hairs on the leaf surface trap humidity in dry climates and create a boundary layer reducing water loss.
- Waxy leaf surfaces reduce water loss.
- Large surface area provides large area for sunlight and shade for plant to minimize heating and reduce water loss.
- In harmful levels of sunlight, specialised leaves, opaque or partly buried, admit light through translucent windows for photosynthesis at inner leaf surfaces (e.g. *Fenestraria*).
- Succulent leaves store water and organic acids for use in CAM photosynthesis.
- Aromatic oils, poisons or pheromones produced by leaf borne glands deter herbivores (e.g. eucalypts).
- Inclusions of crystalline minerals deter herbivores (e.g. silica phytoliths in grasses, raphides in Araceae).

- Petals attract pollinators.
- Spines protect the plants (*e.g.* cacti).
- Special leaves on carnivorous plants are adapted to trapping food, mainly invertebrate prey, though some species trap small vertebrates as well.
- Bulbs store food and water (*e.g.* onions).
- Tendrils allow the plant to climb (*e.g.* peas).
- Bracts and pseudanthia (*false flowers*) replace normal flower structures when the true flowers are greatly reduced (*e.g.* Spurges).
- Spathe.

Interactions with Other Organisms

Although not as nutritious as other organs such as fruit, leaves provide a food source for many organisms. Animals that eat leaves are known as folivores. The leaf is a vital source of energy production for the plant, and plants have evolved protection against folivores such as tannins, chemicals which hinder the digestion of proteins and have an unpleasant taste.

Some species have cryptic adaptations by which they use leaves in avoiding predators. For example, the caterpillars of some leaf-roller moths will create a small home in the leaf by folding it over themselves. Some sawflies similarly roll the leaves of their food plants into tubes. Females of the Attelabidae, so-called leaf-rolling weevils, lay their eggs into leaves that they then roll up as means of protection. Other herbivores and their predators mimic the appearance of the leaf. Reptiles such as some chameleons, and insects such as some katydids, also mimic the oscillating movements of leaves in the wind, moving from side to side or back and forth while evading a possible threat.

Leaf Shape

In botany, leaf shape is characterised with the following terms (botanical Latin terms in brackets):

- Acicular (*acicularis*): Slender and pointed, needle-like
- Acuminate (*acuminata*): Tapering to a long point
- Acute: pointed, having a short sharp apex angled less than 90°
- Aristate (*aristata*): Ending in a stiff, bristle-like point
- Asymmetrical: With the blade shape different on each side of the midrib

- Basal: arising from the crown, bulb, rhizome or corm, etc. as opposed to cauline
- Bipinnate (*bipinnata*): Each leaflet also pinnate
- Caudate: tailed at the apex
- Cauline: borne on the stem as opposed to basal
- Compound: Not simple; the leaf is broken up into separate leaflets, and the leaf blade is not continuous
- Cordate (*cordata*): Heart-shaped, with the petiole or stem attached to the cleft
- Cuneate (*cuneata*): Triangular, stem attaches to point
- Deltoid (*deltoidea*) or deltate: Triangular, stem attaches to side
- Digitate (*digitata*): Divided into finger-like lobes
- Elliptic (*elliptica*): Oval, with a short or no point
- Entire: having a smooth margin without notches or indentations
- Falcate (*falcata*): Sickle-shaped
- Fenestrate (*fenestrata*) "windowed" with holes (e.g. *Monstera deliciosa* or *Aponogeton fenestralis*), or window-like patches of translucent tissue. (cf Perforate)
- Filiform (*filiformis*): Thread- or filament-shaped
- Flabellate (*flabellata*): Semi-circular, or fan-like
- Hastate, spear-shaped (*hastata*): Pointed, with barbs, shaped like a spear point, with flaring pointed lobes at the base
- Laciniate: Very deeply lobed, the lobes being very drawn out, often making the leaf look somewhat like a branch or a pitchfork
- Laminar: Flat (like most leaves)
- Lance-shaped, lanceolate (*lanceolata*): Long, wider in the middle
- Linear (*linearis*): Long and very narrow
- Lobed (*lobata*): With several points
- Mucronate: Ending abruptly in a sharp point
- Obcordate (*obcordata*): Heart-shaped, stem attaches to tapering point
- Oblanceolate (*oblanceolata*): Top wider than bottom
- Oblong (*oblongus*): Having an elongated form with slightly parallel sides
- Obovate (*obovata*): Teardrop-shaped, stem attaches to tapering point

- Obtuse (*obtusus*): With a blunt tip
- Orbicular (*orbicularis*): Circular
- Ovate (*ovata*): Oval, egg-shaped, with a tapering point
- Palmate (*palmata*): Consisting of leaflets or lobes radiating from the base of the leaf.
- Pedate (*pedata*): Palmate, with cleft lobes
- Pedatifid (*pedatifida*)
- Peltate (*peltata*): Rounded, stem underneath
- Perfoliate (*perfoliata*): Stem through the leaves
- Perforate (*perforata*): marked with patches of translucent tissue, as in *Crassula perforata* and *Hypericum perforatum*, or perforated with holes (cf "Fenestrate")
- Pinnate (*pinnata*): Two rows of leaflets
    - o Odd-pinnate, imparipinnate: Pinnate with a terminal leaflet
    - o Paripinnate, even-pinnate: Pinnate lacking a terminal leaflet
    - o Pinnatifid and pinnatipartite: Leaves with pinnate lobes that are not discrete, remaining sufficiently connected to each other that they are not separate leaflets.
    - o Bipinnate, twice-pinnate: The leaflets are themselves pinnately-compound
    - o Tripinnate, thrice-pinnate: The leaflets are themselves bipinnate
    - o Tetrapinnate: The leaflets are themselves tripinnate.
- Pinnatisect (*pinnatifida*): Cut, but not to the midrib (it would be pinnate then)
- Plicate (*plicatus, plicata*): folded into pleats, usually lengthwise, serving the function of stiffening a large leaf.
- Pungent (*spinose*): Having hard, sharp points.
- Reniform (*reniformis*): Kidney-shaped
- Retuse: With a shallow notch in a broad apex
- Rhomboid (*rhomboidalis*): Diamond-shaped
- Round (*rotundifolia*): Circular
- Sagittate (*sagittata*): Arrowhead-shaped
- Simple: Leaf blade in one continuous section, not divided into leaflets (not compound)

- Spear-shaped.
- Spatulate, spathulate (*spathulata*): Spoon-shaped

Figure: Serenoa repens *showing pleated elliptic leaves of seedling and pleated palmate leaves of mature plant*

- Subulate (*subulata*): Awl-shaped with a tapering point
- Subobtuse (*subobtusa*): Somewhat blunted, neither blunt nor sharp
- Sword-shaped (*ensiformis*): Long, thin, pointed
- Trifoliate (or trifoliolate), ternate (*trifoliata*): Divided into three leaflets
- Tripinnate (*tripinnata*): Pinnately compound in which each leaflet is itself bipinnate
- Truncate (*truncata*): With a squared off end
- Unifoliate (*unifoliata*): With a single leaf

Basic Types

Fern: A fern is any one or more of a group of roughly 12,000 species of plants belonging to the botanical group known as Pteridophyta. Unlike mosses, they have xylem and phloem (making them vascular plants). They have stems, leaves, and roots like other vascular plants. Ferns reproduce via spores and have neither seeds nor flowers. Most ferns have what are called fiddleheads. The fiddleheads expand into what are called fronds, which are each delicately divided.

By far the largest group of ferns is the leptosporangiate ferns, but ferns as defined here (also called monilophytes) include horsetails, whisk ferns, marattioid ferns, and ophioglossoid ferns. The term pteridophyte traditionally refers to ferns and a few other seedless vascular plants, although some recent authors have used the term to refer strictly to the monilophytes.

Ferns first appear in the fossil record 360 million years ago in the Devonian Era but many of the current families and species did not appear until roughly 145 million years ago in the early Cretaceous, after flowering plants came to dominate many environments.

Ferns are not of major economic importance, but some are grown or gathered for food, as ornamental plants, for remediating contaminated soils, and have been the subject of research for their ability to remove some chemical pollutants from the air. Some are significant weeds. They also play a role in mythology, medicine, and art.

Life Cycle

Figure: *Gametophyte (thalloid green mass) and sporophyte (ascendent frond) of* Onoclea sensibilis

Ferns are vascular plants differing from lycophytes by having true leaves (megaphylls), which are often pinnate. They differ from seed plants (gymnosperms and angiosperms) in their mode of reproduction—lacking flowers and seeds. Like all other vascular plants, they have a life cycle referred to as alternation of generations, characterized by alternating diploid sporophytic and haploid gametophytic phases. '3

The diploid sporophyte has $2n$ paired chromosomes, where n varies from species to species. The haploid gametophyte has n unpaired chromosomes, i.e. half the number of the sporophyte. The gametophyte of ferns is a free-living organism, whereas the gametophyte of the gymnosperms and angiosperms is dependent on the sporophyte.

Life cycle of a typical fern:

1. A diploid sporophyte phase produces haploid spores by meiosis (a process of cell division which reduces the number of chromosomes by a half).
2. A spore grows into a haploid gametophyte by mitosis (a process of cell division which maintains the number of chromosomes). The gametophyte typically consists of a photosynthetic prothallus.
3. The gametophyte produces gametes (often both sperm and eggs on the same prothallus) by mitosis.
4. A mobile, flagellate sperm fertilizes an egg that remains attached to the prothallus.
5. The fertilized egg is now a diploid zygote and grows by mitosis into a diploid sporophyte (the typical "fern" plant).

Fern Ecology

The stereotypical image of ferns growing in moist shady woodland nooks is far from a complete picture of the habitats where ferns can be found growing. Fern species live in a wide variety of habitats, from remote mountain elevations, to dry desert rock faces, to bodies of water or in open fields. Ferns in general may be thought of as largely being specialists in marginal habitats, often succeeding in places where various environmental factors limit the success of flowering plants.

Some ferns are among the world's most serious weed species, including the bracken fern growing in the Scottish highlands, or the mosquito fern (*Azolla*) growing in tropical lakes, both species forming large aggressively spreading colonies. There are four particular types of habitats that ferns are found in: moist, shady forests; crevices in rock faces, especially when sheltered from the full sun; acid wetlands including bogs and swamps; and tropical trees, where many species are epiphytes (something like a quarter to a third of all fern species).

Figure: *Ferns at Muir Woods, California*

Many ferns depend on associations with mycorrhizal fungi. Many ferns only grow within specific pH ranges; for instance, the climbing fern (*Lygodium palmatum*) of eastern North America will only grow in moist, intensely acid soils, while the bulblet bladder fern (*Cystopteris bulbifera*), with an overlapping range, is only found on limestone.

The spores are rich in lipids, protein and calories, so some vertebrates eat these. The European woodmouse (*Apodemus sylvaticus*) has been found to eat the spores of *Culcita macrocarpa* and the bullfinch (*Pyrrhula murina*) and the New Zealand lesser short-tailed bat (*Mystacina tuberculata*) also eat fern spores.

Fern Structure

Like the sporophytes of seed plants, those of ferns consist of:

- Stems: Fern stems are often referred to as rhizomes, even though they grow underground only in a some of the species. Epiphytic species and many of the terrestrial ones have above-ground creeping stolons (e.g., Polypodiaceae), and many groups have above-ground erect semi-woody trunks (e.g., Cyatheaceae). These can reach up to 20 m tall in a few species (e.g., *Cyathea brownii* on Norfolk Island and *Cyathea medullaris* in New Zealand).

Figure: *Ferns at the Royal Melbourne Botanical Gardens*

Figure: *Tree ferns, probably* Dicksonia antarctica, *growing in Nunniong, Australia*

- Leaf: The green, photosynthetic part of the plant is technically a megaphyll and in ferns, it is often referred to as a frond. New leaves typically expand by the unrolling of a tight spiral called a crozier or fiddlehead fern. This uncurling of the leaf is termed circinate vernation. Leaves are divided into two types:

- o Trophophyll: A vegetative leaf analogous to the typical green leaves of seed plants that does not produce spores, instead only producing sugars by photosynthesis.
- o Sporophyll: A fertile leaf that produces spores borne in sporangia that are usually clustered to form sori. In most ferns, fertile leaves are morphologically very similar to the sterile ones, and they photosynthesize in the same way. In some groups, the fertile leaves are much narrower than the sterile leaves, and may even have no green tissue at all (e.g., Blechnaceae, Lomariopsidaceae).

- Roots: The underground non-photosynthetic structures that take up water and nutrients from soil. They are always fibrous and are structurally very similar to the roots of seed plants.

The gametophytes of ferns, however, are very different from those of seed plants. Instead, they resemble liverworts. A fern gametophyte typically consists of:

- Prothallus: A green, photosynthetic structure that is one cell thick, usually heart or kidney shaped, 3–10 mm long and 2–8 mm broad. The prothallus produces gametes by means of:
  - o Antheridia: Small spherical structures that produce flagellate sperm.
  - o Archegonia: A flask-shaped structure that produces a single egg at the bottom, reached by the sperm by swimming down the neck.
- Rhizoids: root-like structures (not true roots) that consist of single greatly elongated cells, water and mineral salts are absorbed over the whole structure. Rhizoids anchor the prothallus to the soil.

One difference between sporophytes and gametophytes might be summed up by the saying that "Nothing eats ferns, but everything eats gametophytes." This is an over-simplification, but it is true that gametophytes are often difficult to find in the field because they are far more likely to be food than are the sporophytes.

Evolution and Classification

Ferns first appear in the fossil record in the early-Carboniferous period. By the Triassic, the first evidence of ferns related to several modern families appeared. The "great fern radiation" occurred in the late-Cretaceous, when many modern families of ferns first appeared.

One problem with fern classification is the problem of cryptic species. A cryptic species is a species that is morphologically similar to another species, but differs genetically in ways that prevent fertile interbreeding. A good example of this is the currently designated species *Asplenium trichomanes*, the maidenhair spleenwort. This is actually a species complex that includes distinct diploid and tetraploid races. There are minor but unclear morphological differences between the two groups, which prefer distinctly differing habitats. In many cases such as this, the species complexes have been separated into separate species, thus raising the number of overall fern species. Possibly many more cryptic species are yet to be discovered and designated.

Ferns have traditionally been grouped in the Class Filices, but modern classifications assign them their own phylum or division in the plant kingdom, called Pteridophyta, also known as Filicophyta. The group is also referred to as Polypodiophyta, (or Polypodiopsida when treated as a subdivision of tracheophyta (vascular plants), although Polypodiopsida sometimes refers to only the leptosporangiate ferns). The term "pteridophyte" has traditionally been used to describe all seedless vascular plants, making it synonymous with "ferns and fern allies". This can be confusing since members of the fern phylum Pteridophyta are also sometimes referred to as pteridophytes.

Traditionally, three discrete groups of plants have been considered ferns: two groups of eusporangiate ferns—families Ophioglossaceae (adders-tongues, moonworts, and grape-ferns) and Marattiaceae—and the leptosporangiate ferns. The Marattiaceae are a primitive group of tropical ferns with a large, fleshy rhizome, and are now thought to be a sibling taxon to the main group of ferns, the leptosporangiate ferns. Several other groups of plants were considered "fern allies": the clubmosses, spikemosses, and quillworts in the Lycopodiophyta, the whisk ferns in Psilotaceae, and the horsetails in the Equisetaceae. More recent genetic studies have shown that the Lycopodiophyta are more distantly related to other vascular plants, having radiated evolutionarily at the base of the vascular plant clade, while both the whisk ferns and horsetails are as much "true" ferns as are the Ophioglossoids and Marattiaceae. In fact, the whisk ferns and Ophioglossoids are demonstrably a clade, and the horsetails and Marattiaceae are arguably another clade. Molecular data—which remain poorly constrained for many parts of the plants' phylogeny —

have been supplemented by recent morphological observations supporting the inclusion of *Equisetaceae* within the ferns, notably relating to the construction of their sperm, and peculiarities of their roots.

However, there are still differences of opinion about the placement of the Equisetum species. One possible means of treating this situation is to consider only the leptosporangiate ferns as "true" ferns, while considering the other three groups as "fern allies". In practice, numerous classification schemes have been proposed for ferns and fern allies, and there has been little consensus among them.

Their classification based on this phylogeny divides extant ferns into four classes:

- Psilotopsida (whisk ferns and ophioglossoid ferns), about 92 species
- Equisetopsida (horsetails), about 15 species
- Marattiopsida, about 150 species
- Polypodiopsida (leptosporangiate ferns), over 9000 species

The last group includes most plants familiarly known as ferns. Modern research supports older ideas based on morphology that the Osmundaceae diverged early in the evolutionary history of the leptosporangiate ferns; in certain ways this family is intermediate between the eusporangiate ferns and the leptosporangiate ferns. Research by Rai and Graham since this 2006 classification broadly supports the division into four groups, but queries their relationships, concluding that "at present perhaps the best that can be said about all relationships among the major lineages of monilophytes in current studies is that we do not understand them very well".

Uses

Ferns are not as important economically as seed plants but have considerable importance in some societies. Some ferns are used for food, including the fiddleheads of bracken, *Pteridium aquilinum*, ostrich fern, *Matteuccia struthiopteris*, and cinnamon fern, *Osmundastrum cinnamomeum. Diplazium esculentum* is also used by some tropical people as food. Tubers from the King Fern or *para* (*Ptisana salicina*) are a traditional food in New Zealand and the South Pacific. Fern tubers were used for food 30,000 years ago in Europe. Fern tubers were used by the Guanches to make gofio in the Canary Islands. Ferns are generally not known to be poisonous to humans. Licorice fern rhizomes were chewed by the natives of the Pacific Northwest for their flavour.

Ferns of the genus *Azolla* are very small, floating plants that do not resemble ferns. Called mosquito fern, they are used as a biological fertilizer in the rice paddies of southeast Asia, taking advantage of their ability to fix nitrogen from the air into compounds that can then be used by other plants.

Many ferns are grown in horticulture as landscape plants, for cut foliage and as houseplants, especially the Boston fern (*Nephrolepis exaltata*) and other members of the genus Nephrolepis. The Bird's Nest Fern (*Asplenium nidus*) is also popular, as are the staghorn ferns (genus *Platycerium*). Perennial (also known as hardy) ferns planted in gardens in the northern hemisphere also have a considerable following.

Several ferns are noxious weeds or invasive species, including Japanese climbing fern (*Lygodium japonicum*), mosquito fern and sensitive fern (*Onoclea sensibilis*). Giant water fern (*Salvinia molesta*) is one of the world's worst aquatic weeds. The important fossil fuel coal consists of the remains of primitive plants, including ferns.

Ferns have been studied and found to be useful in the removal of heavy metals, especially arsenic, from the soil. Other ferns with some economic significance include:

- *Dryopteris filix-mas* (male fern), used as a vermifuge, and formerly in the US Pharmacopeia; also, this fern accidentally sprouting in a bottle resulted in Nathaniel Bagshaw Ward's 1829 invention of the terrarium or Wardian case
- *Rumohra adiantiformis* (floral fern), extensively used in the florist trade
- *Microsorum pteropus* (Java fern), one of the most popular freshwater aquarium plants.
- *Osmunda regalis* (royal fern) and *Osmunda cinnamomea* (cinnamon fern), the root fiber being used horticulturally; the fiddleheads of *O. cinnamomea* are also used as a cooked vegetable
- *Matteuccia struthiopteris* (ostrich fern), the fiddleheads used as a cooked vegetable in North America
- *Pteridium aquilinum or Pteridium esculentum* (bracken), the fiddleheads used as a cooked vegetable in Japan and are believed to be responsible for the high rate of stomach cancer in Japan. It is also one of the world's most important agricultural weeds, especially in the British highlands, and often poisons cattle and horses.

- *Diplazium esculentum* (vegetable fern), a source of food for some native societies
- *Pteris vittata* (brake fern), used to absorb arsenic from the soil
- *Polypodium glycyrrhiza* (licorice fern), roots chewed for their pleasant flavour
- Tree ferns, used as building material in some tropical areas
- *Cyathea cooperi* (Australian tree fern), an important invasive species in Hawaii
- *Ceratopteris richardii*, a model plant for teaching and research, often called C-fern

Culture

Pteridologist: The study of ferns and other pteridophytes is called pteridology. A pteridologist is a specialist in the study of pteridophytes in a broader sense that includes the more distantly related lycophytes.

Pteridomania

"Pteridomania"' is a term for the Victorian era craze of fern collecting and fern motifs in decorative art including pottery, glass, metals, textiles, wood, printed paper, and sculpture "appearing on everything from christening presents to gravestones and memorials." The fashion for growing ferns indoors led to the development of the Wardian case, a glazed cabinet that would exclude air pollutants and maintain the necessary humidity.The dried form of ferns was also used in other arts, being used as a stencil or directly inked for use in a design. The botanical work, *The Ferns of Great Britain and Ireland,* is a notable example of this type of nature printing. The process, patented by the artist and publisher Henry Bradbury, impressed a specimen on to a soft lead plate. The first publication to demonstrate this was Alois Auer's *The Discovery of the Nature Printing-Process.*

New Zealand Icon

The silver fern in particular has a prominent place within New Zealand culture. Its leaf features as the emblem of many of the country's top national sports teams, including the eponymous Silver Ferns and the All Blacks.

Folklore

Ferns figure in folklore, for example in legends about mythical flowers or seeds. In Slavic folklore, ferns are believed to bloom once a year, during the Ivan Kupala night. Although alleged to be exceedingly difficult to find, anyone who sees a "fern flower" is thought to be guaranteed to be happy and rich for the rest of their life. Similarly, Finnish tradition holds that one who finds the "seed" of a fern in bloom on Midsummer night will, by possession of it, be guided and be able to travel invisibly to the locations where eternally blazing Will o' the wisps called aarnivalkea mark the spot of hidden treasure. These spots are protected by a spell that prevents anyone but the fern-seed holder from ever knowing their locations.

Organisms Confused with Ferns

Misunderstood names:

Several non-fern plants (and even animals) are called "ferns" and are sometimes confused with true ferns. These include:

- "Asparagus fern"—This may apply to one of several species of the monocot genus *Asparagus*, which are flowering plants.
- "Sweetfern"—A flowering shrub of the genus *Comptonia.*
- "Air fern"—A group of animals called hydrozoan that are distantly related to jellyfish and corals. They are harvested, dried, dyed green, and then sold as a "plant" that can "live on air". While it may look like a fern, it is merely the skeleton of this colonial animal..
- "Fern bush"—*Chamaebatiaria millefolium*—a rose family shrub with fern-like leaves.

In addition, the book *Where the Red Fern Grows* has elicited many questions about the mythical "red fern" named in the book. There is no such known plant, although there has been speculation that the oblique grape-fern, *Sceptridium dissectum*, could be referred to here, because it is known to appear on disturbed sites and its fronds may redden over the winter.

Fern-Like Flowering Plants

Some flowering plants such as palms and members of the carrot family have pinnate leaves that somewhat resemble fern fronds. However, these plants have fully developed seeds contained in fruits, rather than the microscopic spores of ferns.

Frond

The term frond refers to a large, divided leaf. In both common usage and botanical nomenclature, the leaves of ferns are referred to as fronds and some botanists restrict the term to this group. Other botanists allow the term frond to also apply to the large leaves of cycads and palms (Arecaceae).

When most people use the word frond they mean a large, compound leaf, but if the term is used botanically to refer to the leaves of ferns, it may be applied to smaller and undivided leaves.

Fronds, like all leaves, usually have a stalk called the petiole supporting a flattened blade, sometimes called a lamina. However, fronds are often described using distinctively different terms. The petiole of a frond is called a stipe and the continuation of the stipe into the blade portion is called the rachis. The blades may be simple (undivided), pinnatifid (deeply incised, but not truly compound), pinnate (compound with the leaflets arranged along a rachis to resemble a feather). If a frond is pinnate, the segments of the blade are called pinnae (singular: pinna) and the stalks bearing the pinnae are called petiolules (The main vein or mid-rib of a pinna is sometimes called a costa (pl., costae).

Figure: *Adaxial (left) and abaxial (right) surfaces of a pinnate fern frond (*Blechnum appendiculatum*). Sori are evident on the abaxial surface.*

If a frond is divided into pinnae, the frond is called once pinnate. In some fronds the pinna are further divided into segments, creating a bipinnate frond. The segments into which each pinna are divided are called pinnules. Rarely, a frond may even be tripinnate, in which case the pinnule divisions are known as ultimate segments.

Pinnae may be arranged along the *rachis* either directly opposite one another or alternating up the stem. The arrangement may change from the base of a blade to the tip, as in the example of *Blechnum* shown below (from base to tip: pinnae opposite to alternate, and pinnatisect to pinnatifid).

Some fronds are not pinnately compound (or simple), but may be palmate or bifurcate. Some ferns, like members of the group Ophioglossales have a unique arrangement.

Fern fronds often bear sporangia, usually on the abaxial surface of the pinnae, but sometimes marginally or scattered over the frond. The sporangia are typically clustered into a sorus (pl., sori). Associated with each sorus in many species is a membranous protective structure called an indusium: an outgrowth of the blade surface that may partly cover the sporangia. Fronds may bear hairs, scales, glands, and, in some species, bulblets for vegetative reproduction.

Fern fronds, as with all leaves, arise from the stem, either directly, or on an outgrowth from the stem termed a phyllopodium. The stem of a typical (leptosporangiate) fern is subterranean or horizontal on the surface of the ground. These stems are called rhizomes. Many fern fronds are initially coiled into a "fiddle-head" or "crozier" although cycad and palm fronds do not have this type of vernation.

Some fern species feature frond dimorphism, in which fertile and sterile fronds differ in appearance and structure.

Pinophyta

Conifer is a Latin word, a compound word of *conus* and *ferre* (to bear), meaning *the one that bears (a) cone(s).*

The conifers, division Pinophyta, also known as division Coniferophyta or Coniferae, are one of 13 or 14 division level taxa within the Kingdom Plantae. Pinophytes are gymnosperms. They are cone-bearing seed plants with vascular tissue; all extant conifers are woody plants, the great majority being trees with just a few being shrubs. Typical examples of conifers include cedars, Douglas-firs, cypresses, firs, junipers, kauri, larches, pines, hemlocks, redwoods, spruces, and yews. The division contains approximately eight families, 68 genera, and 630 living species.

Although the total number of species is relatively small, conifers are of immense ecological importance. They are the dominant plants over huge areas of land, most notably the boreal forests of the northern hemisphere, but also in similar cool climates in mountains further south. Boreal conifers have many wintertime adaptations. The narrow conical shape of northern conifers, and their downward-drooping limbs help them shed snow. Many of them seasonally alter their biochemistry to make them more resistant to freezing, called "hardening". While tropical rainforests have more biodiversity and turnover, the immense conifer forests of the world represent the largest terrestrial carbon sink, i.e. where carbon from atmospheric CO_2 is bound as organic compounds.

They are also of great economic value, primarily for timber and paper production; the wood of conifers is known as softwood.

Evolution

The earliest conifers in the fossil record date to the late Carboniferous (Pennsylvanian) period (about 300 million years ago), possibly arising from *Cordaites*, a seed-bearing plant with cone-like fertile structures. This plant resembled the modern *Araucaria.* Pinophyta, Cycadophyta, and Ginkgophyta all developed at this time. An important adaptation of these gymnosperms was allowing plants to live without being so dependent on water. Other adaptations are pollen (so fertilization can occur without water) and the seed, which lets the embryo be transported and developed elsewhere.

Conifers appear to be one of the taxa that benefited from the Permian–Triassic extinction event.

Taxonomy and Naming

The division name Pinophyta conforms to the rules of the *ICBN*, which state (Article 16.1) that the names of higher taxa in plants (above the rank of family) are either formed from the name of an included family (usually the most common and/or representative), in this case Pinaceae (the pine family), or are descriptive. In the latter case the name for the conifers (at whatever rank is chosen) is Coniferae (Art 16 Ex 2), which is also in widespread use. Older scientific names (no longer allowed) are Coniferophyta and Coniferales.

According to the *ICBN*, it is possible to use a name formed by replacing the termination *-aceae* in the name of an included family, in this case preferably Pinaceae, by the appropriate termination, in the case of this division *-ophyta*. Alternatively, "descriptive botanical names" may also be used at any rank above family. Both are allowed.

This means that if conifers are considered a division, they may be called Pinophyta or Coniferae (as a class they may be called Pinopsida or Coniferae; as an order they may be called Pinales or Coniferae).

Commonly, conifers are considered equivalent to the Gymnosperms, particularly in areas with a temperate climate where they may be the only commonly occurring gymnosperms. However, these are two different levels of grouping: conifers are the largest and economically most important component group of the gymnosperms, but nevertheless they comprise only one of the four groups. The division Pinophyta consists of just one class, Pinopsida, which includes both living and fossil taxa. Subdivision of the living conifers into two or more orders has been proposed from time to time. The most commonly seen in the past was a split into two orders, Taxales (Taxaceae only) and Pinales (the rest), but recent research into DNA sequences suggests that this interpretation leaves the Pinales without Taxales as paraphyletic, and the latter order is no longer considered distinct. A more accurate subdivision would be to split the class into three orders, Pinales containing only Pinaceae, Araucariales containing Araucariaceae and Podocarpaceae, and Cupressales containing the remaining families (including Taxaceae), but there has not been any significant support for such a split, with the majority of opinion preferring retention of all the families within a single order Pinales, despite their antiquity and diverse morphology.

The conifers are now accepted as comprising six to eight families, with a total of 65–70 genera and 600–630 species (696 accepted names). The seven most distinct families are linked in the box above right and phylogenetic diagram left. In other interpretations, the Cephalotaxaceae may be better included within the Taxaceae, and some authors additionally recognise Phyllocladaceae as distinct from Podocarpaceae (in which it is included here). The family Taxodiaceae is here included in family Cupressaceae, but was widely recognised in the past and can still be found in many field guides. A new classification and linear sequence based on molecular data can be found in an article by Christenhusz et al.

The conifers are an ancient group, with a fossil record extending back about 300 million years to the Paleozoic in the late Carboniferous period; even many of the modern genera are recognisable from fossils 60–120 million years old. Other classes and orders, now long extinct, also occur as fossils, particularly from the late Paleozoic and Mesozoic eras. Fossil conifers included many diverse forms, the most dramatically distinct from modern conifers being some herbaceous conifers with no woody stems. Major fossil orders of conifers or conifer-like plants include the Cordaitales, Vojnovskyales, Voltziales and perhaps also the Czekanowskiales (possibly more closely related to the Ginkgophyta).

Morphology

All living conifers are woody plants, and most are trees, the majority having monopodial growth form (a single, straight trunk with side branches) with strong apical dominance. Many conifers have distinctly scented resin, secreted to protect the tree against insect infestation and fungal infection of wounds. Fossilized resin hardens into amber. The size of mature conifers varies from less than one metre, to over 100 metres. The world's tallest, thickest, and oldest living trees are all conifers. The tallest is a Coast Redwood (*Sequoia sempervirens*), with a height of 115.55 metres (although one Victorian mountain ash, *Eucalyptus regnans*, allegedly grew to a height of 140 metres, although the exact dimensions were not confirmed). The thickest, or tree with the greatest trunk diameter, is a Montezuma Cypress (*Taxodium mucronatum*), 11.42 metres in diameter. The smallest is the pygmy pine (*Lepidothamnus laxifolius*) of New Zealand, which is seldom taller than 30 cm tall when mature. The oldest is a Great Basin Bristlecone Pine (*Pinus longaeva*), 4,700 years old. Conflicting sources claim that the largest tree by 3 dimensional volume is either: a Giant Sequoia (*Sequoiadendron giganteum*), with a volume 1486.9 cubic metres or a Ficus benghalensis named Thimmamma Marrimanu with volume unspecified.

Foliage

Since most conifers are evergreens, the leaves of many conifers are long, thin and have a needle-like appearance, but others, including most of the Cupressaceae and some of the Podocarpaceae, have flat, triangular scale-like leaves. Some, notably *Agathis* in Araucariaceae and *Nageia* in Podocarpaceae, have broad, flat strap-shaped leaves. Others such as Araucaria columnaris have leaves that are awl-shaped. In the majority of conifers, the leaves are arranged spirally, exceptions being most of Cupressaceae and one genus in Podocarpaceae, where they are arranged in decussate opposite pairs or whorls of 3 (-4). In many species with spirally arranged leaves, the leaf bases are twisted to present the leaves in a very flat plane for maximum light capture. Leaf size varies from 2 mm in many scale-leaved species, up to 400 mm long in the needles of some pines (e.g. Apache Pine *Pinus engelmannii*). The stomata are in lines or patches on the leaves, and can be closed when it is very dry or cold. The leaves are often dark green in colour, which may help absorb a maximum of energy from weak sunshine at high latitudes or under forest canopy shade. Conifers from hotter areas with high sunlight levels (e.g. Turkish Pine *Pinus brutia*) often have yellower-green leaves, while others (e.g. Blue Spruce *Picea pungens*) have a very strong glaucous wax bloom to reflect ultraviolet light. In the great majority of genera the leaves are evergreen, usually remaining on the plant for several (2-40) years before falling, but five genera (*Larix, Pseudolarix, Glyptostrobus, Metasequoia* and *Taxodium*) are deciduous, shedding the leaves in autumn and leafless through the winter. The seedlings of many conifers, including most of the Cupressaceae, and *Pinus* in Pinaceae, have a distinct juvenile foliage period where the leaves are different, often markedly so, from the typical adult leaves.

Reproduction

Most conifers are monoecious, but some are subdioecious or dioecious; all are wind-pollinated. Conifer seeds develop inside a protective cone called a strobilus. The cones take from four months to three years to reach maturity, and vary in size from 2 mm to 600 mm long.

In Pinaceae, Araucariaceae, Sciadopityaceae and most Cupressaceae, the cones are woody, and when mature the scales usually spread open allowing the seeds to fall out and be dispersed by the wind. In some (e.g. firs and cedars), the cones disintegrate to release the seeds, and in others (e.g. the pines that produce pine nuts) the nut-like seeds are dispersed by birds (mainly nutcrackers, and jays), which break up the specially adapted softer cones. Ripe cones may remain on the plant for a varied amount of time before falling to the ground; in some fire-adapted pines, the seeds may be stored in closed cones for up to 60–80 years, being released only when a fire kills the parent tree.

In the families Podocarpaceae, Cephalotaxaceae, Taxaceae, and one Cupressaceae genus (*Juniperus*), the scales are soft, fleshy, sweet and brightly coloured, and are eaten by fruit-eating birds, which then pass the seeds in their droppings. These fleshy scales are (except in *Juniperus*) known as arils. In some of these conifers (e.g. most Podocarpaceae), the cone consists of several fused scales, while in others (e.g. Taxaceae), the cone is reduced to just one seed scale or (e.g. Cephalotaxaceae) the several scales of a cone develop into individual arils, giving the appearance of a cluster of berries.

The male cones have structures called microsporangia that produce yellowish pollen through meiosis. Pollen is released and carried by the wind to female cones. Pollen grains from living pinophyte species produce pollen tubes, much like those of angiosperms. When a pollen grain lands near a female gametophyte, it undergoes fertilization of the female gametophyte. Alternatively, the gymnosperm male gametophytes are carried by wind to a female cone and are drawn into a tiny opening on the ovule called the micropyle. It is within the ovule that germination occurs. From here, a pollen tube seeks out the female gametophyte and if successful, fertilization occurs. In both cases, the resulting zygote develops into an embryo, which along with its surrounding integument, becomes a seed. Eventually the seed may fall to the ground and, if conditions permit, grows into a new plant.

In forestry, the terminology of flowering plants has commonly though inaccurately been applied to cone-bearing trees as well. The male cone and unfertilized female cone are called *male flower* and *female flower*, respectively. After fertilization, the female cone is termed *fruit*, which undergoes *ripening* (maturation).

Life Cycle

1. To fertilize the ovum, the male cone releases pollen that is carried on the wind to the female cone. (Male and female cones can be found on the same plant)
2. The pollen fertilizes the female gamete (located in the female cone).*
3. A fertilized female gamete (called a zygote) develops into an embryo.
4. Along with integument cells surrounding the embryo, a seed develops containing the embryo. This is an evolutionary characteristic of the gymnosperms.
5. Mature seed drops out of cone onto the ground.
6. Seed germinates and seedling grows into a mature plant.
7. When the plant is mature, the adult plant produces cones and the cycle continues.

Invasive Species

A number of conifers have become invasive species in parts of New Zealand, while *Pinus pinaster*, *Pinus patula* and *Pinus radiata* have become feral in parts of South Africa. These "wilding conifers" are a serious environmental issue causing problems for pastoral farming and for conservation.

Cultivation

Conifers – notably Abies (Fir), Cedrus (Cedar), *Chamaecyparis lawsoniana* (Lawson's cypress), Cupressus (Cypress), Juniper, Picea (Spruce), Pinus (Pine), Taxus (Yew), Thuja - have been the subject of extensive cultivation and hybridisation for ornamental purposes. A multitude of different forms, sizes, and colours are commonly seen in parks and gardens throughout the world.

Lycopodiophyta

The Division Lycopodiophyta (sometimes called Lycophyta or Lycopods) is a tracheophyte subdivision of the Kingdom Plantae. It is the oldest extant (living) vascular plant division at around 410 million years old, and it includes some of the most "primitive" extant species. These species reproduce by shedding spores and have macroscopic alternation of generations, although some are homosporous while others are heterosporous. Members of Lycopodiophyta bear a protostele, and the sporophyte generation is dominant. They differ from all other vascular plants in having microphylls, leaves that have only a single vascular trace (vein) rather than the much more complex megaphylls found in ferns and seed plants.

Classification

There are around 1,200 living (extant) species of Lycopodiophyta which are generally divided into three orders (Lycopodiales, Isoetales, and Selaginellales); in addition there are extinct groups. There is some variation in how the extant orders are grouped into classes: they may be put into a single class; they may be put into two classes, with the Isoetales and Selaginellales combined into one class; or they may be put into three classes, one order in each. The system which uses two classes for extant species is:

- Class Lycopodiopsida – clubmosses and firmosses
- Class Isoetopsida – quillworts, scale trees, and spikemosses
- Class † Zosterophyllopsida – extinct zosterophylls.

Evolution

The members of this division have a long evolutionary history, and fossils are abundant worldwide, especially in coal deposits. In fact, most known genera are extinct. The Silurian species *Baragwanathia longifolia* represents the earliest identifable Lycopodiophyta, while some *Cooksonia* seem to be related. *Lycopodolica* is another Silurian genus which appears to be an early member of this group.

Fossils ascribed to the Lycopodiophyta first appear in the Silurian period, along with a number of other vascular plants. Phylogenetic analysis places them at the base of the vascular plants; they are distinguished by their microphylls and by transverse dehiscence of their sporangia (as contrasted with longitudinal in other vascular plants). Sporangia of living species are borne on the upper surfaces of microphylls (called sporophylls). In some groups, these sporophylls are clustered into strobili.

During the Carboniferous Period, tree-like Lycopodiophyta (such as *Lepidodendron*) formed huge forests that dominated the landscape. The complex ecology of these tropical rainforests collapsed during the mid Pennsylvanian due to a change in climate.

Unlike modern trees, leaves grew out of the entire surface of the trunk and branches, but would fall off as the plant grew, leaving only a small cluster of leaves at the top. Their remains formed many fossil coal deposits. In Fossil Park, Glasgow, Scotland, fossilized Lycopodiophyta trees can be found in sandstone. The trees are marked with diamond-shaped scars where they once had leaves.

Characteristics

Club-mosses are *homosporous*, but spike-mosses and quillworts are *heterosporous*, with female spores larger than the male, and gametophytes forming entirely within the spore walls.

The spores of Lycopodiophyta are highly flammable and so have been used in fireworks. Currently, huperzine, a chemical isolated from a Chinese clubmoss, is under investigation as a possible treatment for Alzheimer's disease.

Monocotyledons, also known as monocots, are one of two major groups of flowering plants (or angiosperms) that are traditionally recognised, the other being dicotyledons, or dicots. Monocot seedlings typically have one cotyledon (seed-leaf), in contrast to the two cotyledons typical of dicots. Monocots have been recognised at various taxonomic ranks, and under various names. The APG III system recognises a clade called "monocots" but does not assign it to a taxonomic rank.

According to the IUCN there are 59,300 species of monocots. The largest family in this group (and in the flowering plants as a whole) by number of species are the orchids (family Orchidaceae), with more than 20,000 species. In agriculture the majority of the biomass produced comes from monocots. The true grasses, family Poaceae (Gramineae), are the most economically important family in this group. These include all the true grains (rice, wheat, maize, etc.), the pasture grasses, sugar cane, and the bamboos. True grasses have evolved to become highly specialised for wind pollination. Grasses produce much smaller flowers, which are gathered in highly visible plumes (inflorescences). Other economically important monocot families are the palm family (Arecaceae), banana family (Musaceae), ginger family (Zingiberaceae) and the amaryllis family (Amaryllidaceae), which includes such ubiquitously used vegetables as onions and garlic.

Many plants cultivated for their blooms are also from the monocot group, notably lilies, daffodils, irises, amaryllis, orchids, cannas, bluebells and tulips.

Name and Characteristics

The name monocotyledons is derived from the traditional botanical name "Monocotyledones", which derives from the fact that most members of this group have one cotyledon, or embryonic leaf, in their seeds. In contrast, the traditional dicotyledons typically have two cotyledons. From a diagnostic point of view the number of cotyledons is neither a particularly useful (as they are only present for a very short period in a plant's life), nor completely reliable characteristic.

Nevertheless, monocots are a distinctive group. One of the most noticeable traits is that a monocot's flower is trimerous, with the flower parts in threes or in multiples of three—having three, six, or nine petals. Many monocots also have leaves with parallel veins.

Figure: Hypoxis decumbens *L. with a typical monocot perigone and parallel leaf venation*

Morphology, Compared to the (Broadly Defined) Dicotyledons

The traditionally listed differences between monocotyledons and dicotyledons are as follows. This is a broad sketch only, not invariably applicable, as there are a number of exceptions. The differences indicated are more true for monocots versus eudicots.

Figure: *Slice of onion, showing parallel veins in cross section*

Figure: Ceroxylon quindiuense *(Quindio wax palm) is considered the tallest monocot in the world*

Feature	*In monocots*	*In dicots*
Number of parts of each flower	in threes (flowers are trimerous)	in fours or fives (tetramerous or pentamerous)
Number of furrows or pores in pollen	one	three
Number of cotyledons (leaves in the seed)	one	two
Arrangement of vascular bundles in the stem	scattered	in concentric circles
Roots	are adventitious	develop from the radicle
Arrangement of major leaf veins	parallel	reticulate

The vast majority of Monocots lack a petiole in their leaves.

A number of these differences are not unique to the monocots. For example, trimerous flowers and monosulcate pollen are also found in magnoliids. Exclusively adventitious roots are found also in Nymphaeaceae and some of the Piperaceae. Similarly, at least one of these traits, parallel leaf veins, is far from universal among the monocots. Monocots with reticulate leaf veins are found in a wide variety of monocot families: for example, *Trillium*, *Smilax* (greenbriar), and *Pogonia* (an orchid), and the Dioscoreales. Nevertheless, this list of traits is a generally valid set of contrasts, especially when contrasting monocots with eudicots rather than non-monocot flowering plants in general.

Emergence

Some monocots, such as grasses, have hypogeal emergence, where the mesocotyl elongates and pushes the coleoptile (which encloses and protects the shoot tip) toward the soil surface. Since elongation occurs above the cotyledon, it is left in place in the soil where it was planted. Many dicots have epigeal emergence, in which the hypocotyl elongates and becomes arched in the soil. As the hypocotyl continues to elongate, it pulls the cotyledons upward, above the soil surface.

Vascular System

Monocots have a distinctive arrangement of vascular tissue known as an atactostele in which the vascular tissue is scattered rather than arranged in concentric rings. Many monocots are herbaceous and do not have the ability to increase the width of a stem (secondary growth) via the same kind of vascular cambium found in non-monocot woody plants. However, some monocots do have secondary growth, and because it does not arise from a single vascular cambium producing xylem inwards and phloem outwards, it is termed "anomalous secondary growth". Examples of large monocots which either exhibit secondary growth, or can reach large sizes without it, are palms (Arecaceae), screwpines (Pandanaceae), bananas (Musaceae), *Yucca*, *Aloe*, *Dracaena*, and *Cordyline*.

Figure: *Stems of two* Roystonea regia *palms showing anomalous secondary growth in monocots. Note the characteristic fibrous roots, typical of monocots.*

Classification

The monocots are considered to form a monophyletic group arising early in the history of the flowering plants. The earliest fossils presumed to be monocot remains date from the early Cretaceous period.

Taxonomists have considerable latitude in naming this group, as the monocots are a group above the rank of family. Article 16 of the *ICBN* allows either a descriptive name or a name formed from the name of an included family.

Historically, the monocotyledons were named:

- Monocotyledoneae in the de Candolle system and the Engler system,
- Monocotyledones in the Bentham & Hooker system and the Wettstein system,
- class Liliopsida in the Takhtajan system and the Cronquist system,
- subclass Liliidae in the Dahlgren system and the Thorne system (1992), and
- clade monocots in the APG system and the APG II system.

Each of these systems uses its own internal taxonomy for the group. The monocotyledons are famous as a group that is extremely stable in its outer borders (it is a well-defined, coherent group), while in its internal taxonomy is extremely unstable (historically no two authoritative systems have agreed with each other on how the monocotyledons are related to each other).

Molecular studies have both confirmed the monophyly of the monocots and helped elucidate relationships within this group. The APG II system does not assign the monocots to a taxonomic rank, instead recognising a monocots clade. This system recognises ten orders of monocots and two families of monocots (Petrosaviaceae and Dasypogonaceae) not yet assigned to any order. More recently, the Petrosaviaceae has been included in the Petrosaviales, and placed near the lilioid orders.

Evolution

For a very long time, fossils of palm trees were believed to be the oldest monocots, first appearing 90 million years ago, but this estimate may not be entirely true (reviewed in Herendeen and Crane, 1995). At least some putative monocot fossils have been found in strata as old as the eudicots (reviewed in Herendeen *et al.*, 1995). The oldest fossils that are unequivocally monocots are pollen from the Late Barremian-Aptian - Early Cretaceous period, about 120-110 million years ago, and are assignable to clade-Pothoideae-Monstereae Araceae; being Araceae, sister to other Alismatales (Friis *et al.*, 2004:) for fossil monocots). They have also found flower fossils of Triuridaceae (Pandanales) in Upper Cretaceous rocks in New Jersey (Gandolfo et al. 2002), becoming the oldest known sighting of saprophytic / mycotrophic habits in angiosperm plants and among the oldest known fossils of monocotyledons.

Topology of the angiosperm phylogenetic tree could infer that the monocots would be among the oldest lineages of angiosperms, which would support the theory that they are just as old as the eudicots. The pollen of the eudicots dates back 125 million years, so the lineage of monocots should be that old too.

Molecular Clock Estimates for the Age of Extant Monocots

Bremer (2000, 2002), using rbcL sequences and the mean path length method ("mean-path lengths method"), estimated the age of the monocot crown group (i.e., the time at which the ancestor of today's *Acorus* diverged from the rest of the group) as 134 million years. Similarly, Wikström et al. (2001), using Sanderson's (1997) non-parametric rate smoothing approach ("nonparametric rate smoothing approach"), obtained ages of 158 or 141 million years for the crown group of monocots. All these estimates have large error ranges (usually 15-20%), and Wikström et al. used only a single calibration point, namely the split between Fagales and Cucurbitales, which was set to 84 Ma, in the late Santonian period). Early molecular clock studies using strict clock models had estimated the monocot crown age to 200 ± 20 million years ago (Savard et al. 1994) or 160 ± 16 million years (Goremykin et al. 1997), while studies using relaxed clocks have obtained 135-131 million years (Leebens-Mack et al. 2005) or 133.8 to 124 million years (Moore et al. 2007). Bremer's estimate (2000) of 134 million years has been used as a secondary calibration point in other analyses (Janssen and Bremer 2004).

Core Group

The age of the core group of so-called 'nuclear monocot' or 'core monocots' by the Angiosperm Phylogeny Website ("core monocots" in English), which correspond to all orders except Acorales and Alismatales, is about 131 million years to present, and crown group age is about 126 million years to the present. The subsequent branching in this part of the tree (i.e., Petrosaviaceae, Dioscoreales + Pandanales and Liliales clades appeared), including the crown Petrosaviaceae group may be in the period around 125-120 million years BC (about 111 million years so far in Bremer 2000), and stem groups of all other orders, including Commelinadae would have diverged about or shortly after 115 million years (Janssen and Bremer 2004). These and many clades within these orders may have originated in southern Gondwana, i.e., Antarctica, Australasia, and southern South America (Bremer and Jansen 2006).

Aquatic Monocots

The aquatic monocot Alismatales have commonly been regarded as "primitive" (Hallier, 1905, Arber 1925, Hutchinson, 1934, Cronquist 1968, 1981, Takhtajan 1969, 1991 Stebbins 1974, Thorne 1976). They have also been considered to have the most primitive foliage, which were cross-linked as Dioscoreales (Dahlgren et al.). 1985 and Melanthiales (Thorne 1992a, 1992b). Keep in mind that, as stressed by Soltis et al. 2005, the "most primitive" monocot is not necessarily "the sister of everyone else." This is because the ancestral or primitive characters are inferred by means of the reconstruction of characteristic states, with the help of the phylogenetic tree. So primitive characters of monocots may be present in some derived groups. On the other hand, the basal taxa may exhibit many morphological autapomorphies.

So although Acoraceae is the sister group to the remaining monocotyledons, the result does not imply that Acoraceae is "the most primitive monocot" in terms of its characteristics. In fact, Acoraceae is highly derived in most morphological characteristics, which is precisely why so many Alismatales Acoraceae occupied relatively imitative positions in trees produced by Chase et al. 1995b and Stevenson and Loconte 1995.

Some authors support the idea of an aquatic phase as the origin of monocots (Henslow 1893, and also cited and argued in the phylogeny section that Alismatales are the most primitive). The phylogenetic position of Alismatales (many water), which occupy a relationship with the rest except the Acoraceae, do not rule out the idea, because it could be 'the most primitive monocots' but not 'the most basal'.

The Atactostele stem, the long and linear leaves, the absence of secondary growth (see the biomechanics of living in the water), roots in groups instead of a single root branching (related to the nature of the substrate), including sympodial use, are consistent with a water source. However, while monocots were sisters of the aquatic Ceratophyllales, or their origin is related to the adoption of some form of aquatic habit, it would not help much to the understanding of how it evolved to develop their distinctive anatomical features: the monocots seem so different from the rest of angiosperms and it's difficult to relate their morphology, anatomy and development and those of broad-leaved angiosperms (e.g. Zimmermann and Tomlinson 1972; Tomlinson 1995).

Other Taxa

In the past, taxa which had petiolate leaves with reticulate venation were considered "primitive" within the monocots, because of its superficial resemblance to the leaves of dicotyledons. Recent work suggests that these taxa are sparse in the phylogenetic tree of monocots, such as fleshy fruited taxa (excluding taxa with aril seeds dispersed by ants), the two features would be adapted to conditions that evolved together regardless (Dahlgren and Clifford 1982; Patterson and Givnish 2002, Givnish et al. 2005b, 2006b). Among the taxa involved were *Smilax*, *Trillium* (Liliales), *Dioscorea* (Dioscoreales), etc.

A number of these plants are vines that tend to live in shaded habitats for at least part of their lives, and may also have a relationship with their shapeless stomata. Reticulate venation seems to have appeared at least 26 times in monocots, in fleshy fruits 21 times (sometimes lost later), and the two characteristics, though different, showed strong signs of a tendency to be good or bad in tandem, a phenomenon Givnish et al. (2005b, 2006b) described as "concerted convergence" ("coordinated convergence").

Succulent Plant

In botany, succulent plants, also known as succulents or sometimes fat plants, are plants having some parts that are more than normally thickened and fleshy, usually to retain water in arid climates or soil conditions. Succulent plants may store water in various structures, such as leaves and stems. Some definitions also include roots, so that geophytes that survive unfavourable periods by dying back to underground storage organs may be regarded as succulents. In horticultural use, the term "succulent" is often used in a way which excludes plants that botanists would regard as succulents, such as cacti. Succulents are grown as ornamental plants because of their striking and unusual appearance.

Definition

There are a number of somewhat different definitions of the term *succulent*. One difference lies in whether or not roots are included in the parts of a plant which make it a succulent. Some authors include roots, as in the definition "plants in which the leaves, stem or roots have become more than usually fleshy by the development of water-storing tissue." Others exclude roots, as in the definition "a plant with thick, fleshy and swollen stems and/or leaves, adapted to dry environments". This difference affects the relationship between succulents and "geophytes" – plants that survive unfavourable seasons as a resting bud on an underground organ. These underground organs, such as bulbs, corms and tubers, are often fleshy with water-storing tissues. Thus if roots are included in the definition, many geophytes would be classed as succulents.

Plants adapted to living in dry environments are termed *xerophytes*; thus succulents are often xerophytes. However, not all xerophytes are succulents, since there are other ways of adapting to a shortage of water, e.g. by developing small leaves which may roll up or having leathery rather than succulent leaves. Nor are all succulents xerophytes, since plants like *Crassula helmsii* are both succulent and aquatic.

Those who grow succulents as a hobby use the term in a different way to botanists. In horticultural use, the term *succulent* regularly excludes cacti. For example, Jacobsen's three volume *Handbook of Succulent Plants* does not cover cacti, and "cacti and succulents" is the title or part of the title of many books covering the cultivation of these plants. However, in botanical terminology, cacti are succulents. Horticulturists may also exclude other groups of plants, e.g. bromeliads. A practical, but unscientific, horticultural definition is "a succulent plant is any desert plant that a succulent plant collector wishes to grow". Such plants less often include geophytes (in which the swollen storage organ is wholly underground) but do include plants with a caudex, which is a swollen above-ground organ at soil level, formed from a stem, a root or both.

A further difficulty is that plants are not either *succulent* or *non-succulent*. In many genera and families there is a continuous sequence from plants with thin leaves and normal stems to those with very clearly thickened and fleshy leaves or stems, so that deciding what is a succulent is often arbitrary. Different sources may classify the same plant differently.

Appearance

The storage of water often gives succulent plants a more swollen or fleshy appearance than other plants, a characteristic known as succulence. In addition to succulence, succulent plants variously have other water-saving features. These may include:

- Crassulacean acid metabolism (CAM) to minimize water loss
- absent, reduced, or cylindrical-to-spherical leaves
- reduction in the number of stomata
- stems as the main site of photosynthesis, rather than leaves
- compact, reduced, cushion-like, columnar, or spherical growth form
- ribs enabling rapid increases in plant volume and decreasing surface area exposed to the sun
- waxy, hairy, or spiny outer surface to create a humid micro-habitat around the plant, which reduces air movement near the surface of the plant, and thereby reduces water loss and creates shade
- roots very near the surface of the soil, so they are able to take up moisture from very small showers or even from heavy dew
- ability to remain plump and full of water even with high internal temperatures (e.g. 52 °C or 126 °F)
- very impervious outer cuticle (skin)
- mucilaginous substances, which retain water abundantly

Habitat

Many succulents come from the dry areas of the tropics and subtropics, such as steppes, semi-desert, and desert. High temperatures and low precipitation force plants to collect and store water to survive long dry periods. Succulents may occasionally occur as epiphytes - "air plants" - as they have limited or no contact with the ground, and are dependent on their ability to store water and gain nutrients by other means; this niche is seen in *Tillandsia*. Succulents also occur as inhabitants of sea coasts and dry lakes, which are exposed to high levels of dissolved minerals that are deadly to many other plant species.

Evolution

The best-known succulents are cacti (family: Cactaceae). Virtually all cacti are succulents, but not all succulents are cacti. A unique feature of cacti is the possession of areoles, structures from which spines and flowers are produced.

To differentiate between these two basic types that seem so similar, but that are unrelated succulent plants, use of the terms, *cactus* or *cacti*, only should be used to describe succulents in the cactus family. Popular collection of these types of plants has led to many Old World plants becoming established in the wild in the New World, and vice versa.

Cataphyll

In plant morphology, a cataphyll (sometimes also called a cataphylla, or cataphyll leaf) is a leaf whose primary function is something other than photosynthesis. Cataphylls are at most trivially or transiently photosynthetic, and instead of photosynthesis the main functions of most types are storage, protection, or structural support. Many forms of cataphylls die in performing their function. Cataphylls such as bud scales often are shed after the need for them is past, but dead cataphylls that afford protection from weather or pests may be accumulated into long-lasting, thick coverings.

Forms of Cataphylls

Some kinds of cataphylls perform a transient function, after which they die and may be shed. Those that are shed early are said to be caducous, but that term can apply to any organ that is shed early, not only leaves; for example, many *Geraniums* have caducous stamens. The sepals of *Papaver* species are shed during the very opening of the petals, and as such they are a dramatic example of caducous leaves. Many other forms of cataphylls, such as some spines, are persistent, but cannot perform their major function until they die, whether they physically get shed or not. Yet others perform indefinitely in the form of persistent structures that remain on the plant after they die. Examples of various kinds of cataphylls include bud-scales, bulb-scales, corm-scales, rhizome-scales, cotyledons, scaly bracts, spines and perhaps glochids. Each of these occurs in various forms and contexts; for example, bud-scales occur on various kinds of leaf or branch buds as well as on flower buds.

At all events, cataphylls are in general *sacrificial organs*; their function is not related to their own survival, but in direct or indirect support of the propagation of the parent organism.

The word *cataphyll* derives from the Greek; in context it means something like “leaf to be broken down”, implying leaves that are discarded or consumed. In fact some forms of cataphylls, such as the leaves or leaf bases forming the tunic around a corm or a bulb, are retained after they have died, and the protective presence of their remains is their major function. Similar protective masses of dead leaves encircle the stems of some species of palm trees or *aloes*, but those are not usually regarded as cataphylls because their primary function while alive was photosynthesis, as is usual for leaves. Clearly, the precise limits of the definition of the concept is a matter of convenience in any given context.

Cotyledons as Cataphylls

Cotyledons are widely regarded as a class of cataphyll, though many kinds of cotyledon function as living tissue and remain alive till the end of their function at least, at which time they wither and may drop off. They begin as leaf rudiments and many kinds accumulate nutrient materials for storage, starting to give up their stored material as the plant begins to germinate. Some, such as the cotyledons of many legumes, conifers, and cucurbits, even develop chlorophyll and perform the first photosynthesis for the germinating plant. Logically it is stretching the term to call such an organ a cataphyll, because after all, most ordinary leaves also have limited life spans and drop off when exhausted, whether after one season or several. However, non-cotyledonous leaves are not normally regarded as cataphylls just because they do not live as long as the parent plant. If non-permanent leaves were termed cataphylls, then few leaves would be anything but cataphylls; the most convincing example of non-cataphylls then might be the two persistent leaves of a *Welwitschia* plant. Those two leaves of the *Welwitschia* must last the plant for its entire life span, typically many centuries, because the growth point, the apical meristem, dies early in the development of the seedling. Interestingly, apart from those two main leaves, the two cotyledons of *Welwitschia* also are persistent in a sense; their remnants form the basis of the obconical, sideways, growth of the mouth-like slit at the top of the stem.

Such special examples aside, most cotyledons are as it were, disposable once they have performed their transient function, with or without photosynthesis, and in suitable contexts may be regarded as cataphylls. It would not be practical to demand a sharply-distinct definition for a continuous range of widely-varied functions.

Spines as Cataphylls

It also is a matter of context and preference whether one regards any particular kind of spine as a cataphyll or not. The terminology for glochids in particular is confusing, as they are variously and arbitrarily referred to as spines, bristles and more. Morphologically only *spines* could strictly speaking be cataphylls, because the others are not leaves, but in the current context the point is hardly worth pursuing.

Some kinds of such defencive organs remain on the plant, whereas others, such as the glochids of *Cacti* in the sub-family *Opuntioideae*, not only detach when touched, but owe much of their very function to their tendency to remain stuck into the skin, respiratory system, or eyes of the victim. In either case, most of the tissue in most kinds of spines will be dead by the time the spine is ready for action, whereas at the start of its development a spine generally is soft and fleshy, alive, photosynthesising, and growing, but ineffective for defence. Spines and glochidia also are sacrificial organs in a sense.

Buds, Flowers, and Associated Cataphylls

Bud-scales and bract-scales (or scaly bracts) are leaves that have a specific protective function with at most trivial and transient photosynthetic function; they are vital protection against pests and climate, especially during periods of dormancy. Their most spectacularly specialised examples are the often precisely imbricate bud scales of the broad-leaved trees of boreal forests.

In warm temperate climates one also finds some plants with armored buds, even if there is no winter dormancy. For example, many species of sumach indigenous to temperate zones, such as (*Searsia* or *Rhus*), have naked apical buds that continue growing throughout the active season and never go into a state of dormancy before their stems lapse into senescence. However, each leaf on a shoot has an axillary bud, and not every axillary bud begins to grow as soon as it is formed; instead it might take years before there is occasion for such an axillary bud to open; it might never open at all. For protection during its dormancy, if any, the first few leaves of each axillary bud grow into a snug imbricate covering of cataphyllic bud-scales soon after the bud forms. In this article, pictures of *Searsia angustifolia* illustrate the effect.

There are yet other classes of bud protection; large leaves of tropical plants without any dormant stage to speak of, such as *Philodendrons*, often develop within an unusually large protective cataphyll, possibly functioning largely as scaffolding for the growing leaf, or protection from wind during the period when their tissue is tender and their fibres are undeveloped. The protective cataphylls curl back and dry out as the leaf opens and matures, after which they often are shed. The protective bracts growing in and around the inflorescences of *Musa* species, such as bananas, amount to cataphylls protecting their flowers and young fruit. They die and may be shed as the inflorescence matures.

Figure: *Calyx of ripe Cape gooseberry, sepals on the ripe fruit turn papery and fragile after their protective function lapses.*

One also could argue for regarding the sepals of some plants as cataphylls. Consider two contrasting examples — the sepals of the "Cape gooseberry" unite to form a persistent protective shroud around the fruit, becoming papery and brittle as the fruit ripens, though the cover is not actually shed. Conversely the sepals of most poppies are strictly caducous; after having protected the flower bud, they are physically shed during the process of the opening of the petals.

Bulb-Scales

Bulb scales, such as those comprising the bulb of an onion or *Amaryllis*, are cataphylls primarily in that part of the organ typically never acts as photosynthesising leaf tissue. However, that is something like special pleading, because many or most of the bulb scales were at one time just the non-photosynthesising leaf bases of photosynthesising, apparently deciduous leaves.

Figure: *Onions with cross-section, showing roots, scales, and tunic layers (onionskin).*

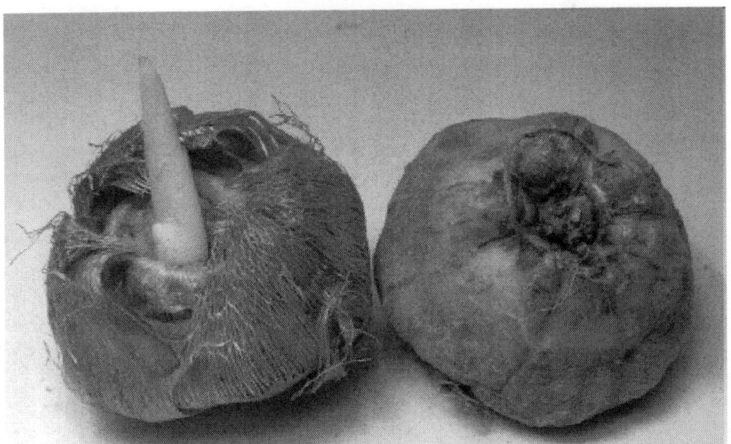

Figure: *Ixia corms, one sprouting and showing tunic layers, the other with tunics removed, inverted to show the root base. The stem is solid, not made up of leaves like those making up the onion bulb.*

It is in fact possible for most bulb scales to develop photosynthetic tissue, and some plants adapted to arid, sun-baked conditions on stony soils, such as some Karooid species of Ornithogalum, commonly lie on the soil surface, with the exposed bulbs as significant photosynthetic organs. However, most bulbs are subterranean, with scales that never see the light. In such species only the visible parts of the leaves are green and deciduous, and the bulb scales stay in place as living, non-photosynthetic storage organs.

On most bulbs some scales are the bases of leaves that never were photosynthetic, but immediately formed cataphylls that function either as storage organs or as protective scales, or both in turn. New leaves in true bulbs, botanically speaking, are produced from the centre, and as they grow, they force the older bulb scales outwards.

The outermost scales of the plant yield up their stores to the plant each season and as their depleted tissue dies, the residue gets added to the bulb's protective tunic. Only then does the dying leaf complete its role as a cataphyll in the usual sense of dying and being discarded as it breaks down. Most onions live for only a few years, but for example undamaged Amaryllis or Boophone bulbs in the wild can grow indefinitely, and each new bulb scale can take many years to reach the outside of the bulb and merge into the tunic.

Corm-Scales

Like bulb-scales, corm-scales are largely the basal parts of the photosynthetic leaves that show above ground.

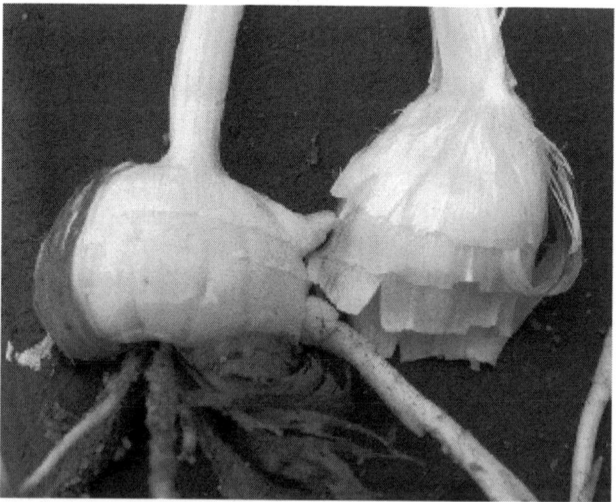

Figure: *Crocosmia corm with tunic stripped partly off to show its constitution of the basal parts of leaves arising from nodes on the corm. Such leaves, especially early leaves that never performed much photosynthesis, amount to true cataphylls*

Some species of cormous plants, such as some *Lapeirousias* also produce cataphyllous leaves that act as practically nothing more than tunic leaves for the corm. Unlike bulb-scales however, the corm tunic has no significant storage function; that task is left to the parenchyma of the cortex of the corm.

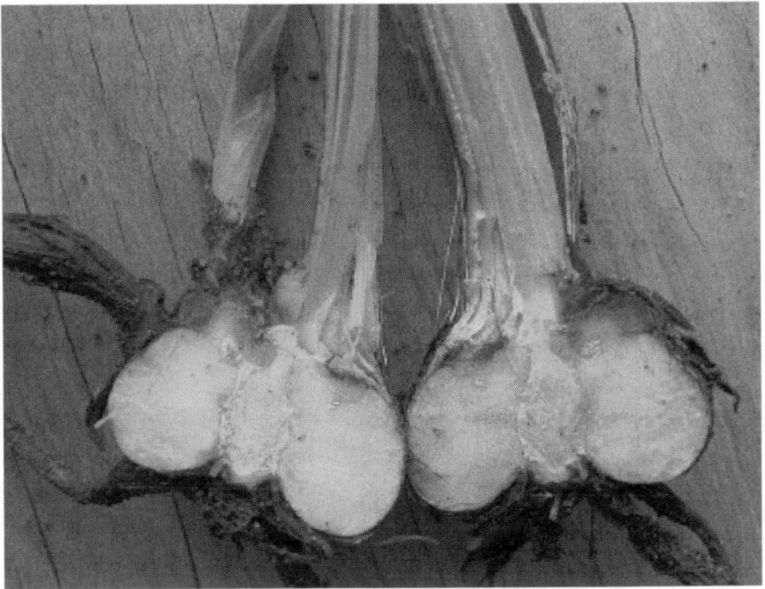

Figure: *Corm of Crocosmia, split to show tunic and leaves, as well as a new corm growing from a bud on the cortex of the old corm*

2

Seeking Illumination

Autumn Leaf Colour

Autumn leaf colour is a phenomenon that affects the normally green leaves of many deciduous trees and shrubs by which they take on, during a few weeks in the autumn season, various shades of red, yellow, purple, and brown. The phenomenon is commonly called autumn colours or autumn foliage in British English and fall colours, fall foliage, or simply foliage in American English.

In some areas of Canada and the United States, "leaf peeping" tourism is a major contribution to economic activity. This tourist activity occurs between the beginning of colour changes and the onset of leaf fall.

Chlorophyll and the Green Colour

A green leaf is green because of the presence of a pigment known as chlorophyll, which is inside an organelle called a chloroplast. When they are abundant in the leaf's cells, as they are during the growing season, the chlorophylls' green colour dominates and masks out the colours of any other pigments that may be present in the leaf. Thus the leaves of summer are characteristically green.

Chlorophyll has a vital function: that of capturing solar rays and utilising the resulting energy in the manufacture of the plant's food—simple sugars which are produced from water and carbon dioxide.

These sugars are the basis of the plant's nourishment—the sole source of the carbohydrates needed for growth and development. In their food-manufacturing process, the chlorophylls themselves break down and thus are being continually "used up". During the growing season, however, the plant replenishes the chlorophyll so that the supply remains high and the leaves stay green.

Figure: *In this leaf, the veins are still green while the other tissue is turning red. This produces a fractal-like pattern*

In late summer, as daylight hours shorten and temperatures cool, the veins that carry fluids into and out of the leaf are gradually closed off as a layer of special cork cells forms at the base of each leaf. As this cork layer develops, water and mineral intake into the leaf is reduced, slowly at first, and then more rapidly. It is during this time that the chlorophyll begins to decrease.

Often the veins will still be green after the tissues between them have almost completely changed colour.

A lot of chlorophyll is located in Photosystem II (Light Harvesting Complex II or LHC II), the most abundant membrane protein on earth. LHC II is where light is captured in photosynthesis. It is located in the thylakoid membrane of the chloroplast and it is composed of an apoprotein along with several ligands, the most important of which are chlorophylls a and b. In the fall, this complex is broken down. Chlorophyll degradation is thought to occur first. Recent research suggests that the beginning of chlorophyll degradation is catalyzed by chlorophyll b reductase, which reduces chlorophyll b to 7 hydroxymethyl chlorophyll a, which is then reduced to chlorophyll a. This is believed to destabilize the complex, at which point breakdown of the apoprotein occurs. An important enzyme in the breakdown of the apoprotein is FtsH6, which belongs to the FtsH family of proteases. Chlorophylls degrade into colourless tetrapyrroles known as nonfluorescent chlorophyll catabolites (NCCs). As the chlorophylls degrade, the hidden pigments of yellow xanthophylls and orange beta-carotene are revealed. These pigments are present throughout the year, but the red pigments, the anthocyanins, are synthesized *de novo* once roughly half of chlorophyll has been degraded. The amino acids released from degradation of light harvesting complexes are stored all winter in the tree's roots, branches, stems, and trunk until next spring when they are recycled to re leaf the tree.

Pigments that Contribute to Other Colours

Carotenoids

Carotenoids are present in leaves the whole year round, but their orange-yellow colours are usually masked by green chlorophyll. As autumn approaches, certain influences both inside and outside the plant cause the chlorophylls to be replaced at a slower rate than they are being used up. During this period, with the total supply of chlorophylls gradually dwindling, the "masking" effect slowly fades away. Then other pigments that have been present (along with the chlorophylls) in the cells all during the leaf's life begin to show through. These are carotenoids and they provide colourations of yellow, brown, orange, and the many hues in between.

The carotenoids occur, along with the chlorophyll pigments, in tiny structures called plastids within the cells of leaves. Sometimes they are in such abundance in the leaf that they give a plant a yellow-green colour, even during the summer. Usually, however, they become prominent for the first time in autumn, when the leaves begin to lose their chlorophyll.

Carotenoids are common in many living things, giving characteristic colour to carrots, corn, canaries, and daffodils, as well as egg yolks, rutabagas, buttercups, and bananas.

Their brilliant yellows and oranges tint the leaves of such hardwood species as hickories, ash, maple, yellow poplar, aspen, birch, black cherry, sycamore, cottonwood, sassafras, and alder. Carotenoids are the dominant pigment in colouration of about 15-30% of tree species.

Anthocyanins

The reds, the purples, and their blended combinations that decorate autumn foliage come from another group of pigments in the cells called anthocyanins. Unlike the carotenoids, these pigments are not present in the leaf throughout the growing season, but are actively produced towards the end of summer. They develop in late summer in the sap of the cells of the leaf, and this development is the result of complex interactions of many influences — both inside and outside the plant. Their formation depends on the breakdown of sugars in the presence of bright light as the level of phosphate in the leaf is reduced.

Figure: *English country lane in Autumn*

During the summer growing season, phosphate is at a high level. It has a vital role in the breakdown of the sugars manufactured by chlorophyll. But in the fall, phosphate, along with the other chemicals and nutrients, moves out of the leaf into the stem of the plant. When this happens, the sugar-breakdown process changes, leading to the production of anthocyanin pigments. The brighter the light during this period, the greater the production of anthocyanins and the more brilliant the resulting colour display. When the days of autumn are bright and cool, and the nights are chilly but not freezing, the brightest colourations usually develop.

Anthocyanins temporarily colour the edges of some of the very young leaves as they unfold from the buds in early spring. They also give the familiar colour to such common fruits as cranberries, red apples, blueberries, cherries, strawberries, and plums.

Anthocyanins are present in about 10% of tree species in temperate regions, although in certain areas — most famously New England — up to 70% of tree species may produce the pigment. In autumn forests they appear vivid in the maples, oaks, sourwood, sweetgums, dogwoods, tupelos, cherry trees and persimmons. These same pigments often combine with the carotenoids' colours to create the deeper orange, fiery reds, and bronzes typical of many hardwood species.

Cell Walls

The brown colour of leaves is not the result of a pigment, but rather cell walls, which may be evident when no colouring pigment is visible.

Function of Autumn Colours

Deciduous plants were traditionally believed to shed their leaves in autumn primarily because the high costs involved in their maintenance would outweigh the benefits from photosynthesis during the winter period of low light availability and cold temperatures. In many cases this turned out to be over-simplistic — other factors involved include insect predation, water loss, and damage from high winds or snowfall.

Anthocyanins, responsible for red-purple colouration, are actively produced in autumn, but not involved in leaf-drop. A number of hypotheses on the role of pigment production in leaf-drop have been proposed, and generally fall into two categories: interaction with animals, and protection from non-biological factors.

Photoprotection

According to the photoprotection theory, anthocyanins protects the leaf against the harmful effects of light at low temperatures. It is true that the leaves are about to fall and therefore it is not of extreme importance for the tree to protect them. Photo-oxidation and photo-inhibition, however, especially at low temperatures, make the process of reabsorbing nutrients less efficient. By shielding the leaf with anthocyanins, according to the photoprotection theory, the tree manages to reabsorb nutrients (especially nitrogen) more efficiently.

Coevolution

According to the coevolution theory, the colours are warning signals towards insects that use the trees as a host for the winter, for example aphids. If the colours are linked to the amount of chemical defences against insects, then the insects will avoid red leaves and increase their fitness; at the same time trees with red leaves will have an advantage because they reduce their parasite load. This has been shown in the case of apple trees where some but not all domesticated apple varieties unlike wild ones lack red leaves in autumn. A greater proportion of aphids that avoid apple trees with red leafs manage to grow and develop compared to those that do not. A trade off moreover exists between fruit size, leaf colour and aphids resistance as varieties with red leaves have smaller fruits suggesting a cost to the production of red leafs linked to a greater need for reduced aphid infestation. Consistent with red leaved tree providing reduced survival for aphids, tree species with bright leaves tend to select for more specialist aphid pests than do trees lacking bright leaves (autumn colours are useful only in those species coevolving with insect pests in autumn).

The coevolution theory of autumn colours was proposed by W. D. Hamilton in 2001 as an example of evolutionary signalling theory. Biological signals such a red leafs it is argued because they are costly to produce are usually honest and so signal the true quality of the signaller with low quality individuals being unable to fake them and so cheat. Autumn colours would be a signal if they are costly to produce, or be impossible to fake (for example if autumn pigments were produced by the same biochemical pathway that produces the chemical defences against the insects).

The change of leaf colours prior to fall have also been suggested as adaptations that may help to undermine the camouflage of herbivores. Many plants with berries attract birds with especially visible berry and/or leaf colour, particularly bright red. The birds get a meal while the shrub, vine or typically small tree gets undigested seeds carried off and deposited with the birds' manure. Poison Ivy is particularly notable for having bright red foliage drawing birds to its off-white seeds (which are edible for birds, but not most mammals).

Allelopathy

Researchers at New York's Colgate University have found evidence that the brilliant red colours of maple leaves is created by a separate processes than those in chlorophyll breakdown. At the very time when the tree is struggling to cope with the energy demands of a changing and challenging season maple trees are involved in an additional metabolic expenditure to create anthocyanins. These anthocyanins, which create the visual red hues, have been found to aid in interspecific competition by stunting the growth of nearby saplings in what is known as allelopathy. (Frey & Eldridge, 2005)

Tourism

Although some autumn colouration occurs wherever deciduous trees are found, the most brightly coloured autumn foliage is found in four or five regions of the world: most of southern mainland Canada; most of the eastern part of the United States as well as smaller areas of forest further west; Scandinavia, Northern, and Western Europe north of the Alps; the Caucasus region near the Black Sea, Russia and Eastern Asia, including much of northern and eastern China, as well as Argentina, Chile, southern Brazil, Korea, Japan and New Zealand's South Island.

Eastern Canada and the New England region of the United States are famous around the world for the brilliance of their "fall foliage," and a seasonal tourist industry has grown up around the few weeks in autumn when the leaves are at their peak. Thick forest cover and distinct seasonal changes make this part of the world an ideal setting for the types of deciduous trees that produce wonderful fall foliage. Fall colours are typically at their peaks in early to mid October for much of the northern and interior parts of the area, late October for areas further south, and early November for the warmer subtropical areas of the region. Some television and web-based weather forecasts even report on the status of the fall foliage throughout the season as a service to tourists, the most well-known of which is The Weather Channel. Fall foliage tourists are often referred to as "leaf peepers". Fall foliage tours to the Rocky Mountain states, the northwestern United States and far western Canada are becoming more popular as well. The Japanese *momijigari* tradition is similar, though more closely related to *hanami*. In Finland the time of the year is called *ruska* (russeting). In Latvia during russeting, leaf peeping is promoted both internationally and locally. Other lands may also promote this, but mushrooming may be more culturally significant, for example in Lithuania.

Climate Influences

Compared to Western Europe, North America provides many more arbor species (more than 800 species and about 70 oaks, compared to 51 and three respectively in Western Europe) which adds many more different colours to the spectacle. The main reason was the different effect of the ice ages—while in North America, species were protected in more southern regions along north–south ranging mountains, which was not the case in Europe.

Global warming and rising carbon dioxide levels in the atmosphere delay the usual autumn spectacle of changing colours and falling leaves in northern hardwood forests, and increase forest productivity. Experiments with poplar trees showed that they stayed greener longer with higher CO_2 levels, independent of temperature changes. However, the experiments over two years were too brief to indicate how mature forests may be impacted over time. Also, other factors, such as increasing ozone levels close to the ground (tropospheric ozone pollution), can negate the beneficial effects of elevated carbon dioxide.

Deciduous

Deciduous means "falling off at maturity" or "tending to fall off", and it is typically used in reference to trees or shrubs that lose their leaves seasonally and to the shedding of other plant structures such as petals after flowering or fruit when ripe. In a more general sense, *deciduous* means "the dropping of a part that is no longer needed" or "falling away after its purpose is finished". In plants it is the result of natural processes. "Deciduous" has a similar meaning when referring to animal parts, such as deciduous antlers in deer or deciduous teeth, also known as baby teeth, in some mammals (including humans).

Botany

In botany and horticulture, deciduous plants, including trees, shrubs and herbaceous perennials, are those that lose all of their leaves for part of the year. This process is called abscission. In some cases leaf loss coincides with winter — namely in temperate or polar climates. In other parts of the world, including tropical, subtropical, and arid regions, plants lose their leaves during the dry season or other seasons, depending on variations in rainfall.

The converse of deciduous is evergreen, where foliage is shed on a different schedule from deciduous trees, therefore appearing to remain green year round. Plants that are intermediate may be called semi-deciduous; they lose old foliage as new growth begins. Other plants are semi-evergreen and lose their leaves before the next growing season, retaining some during winter or dry periods. Some trees, including a few species of oak, have desiccated leaves that remain on the tree through winter; these dry persistent leaves are called marcescent leaves and are dropped in the spring as new growth begins.

Many deciduous plants flower during the period when they are leafless, as this increases the effectiveness of pollination. The absence of leaves improves wind transmission of pollen for wind-pollinated plants and increases the visibility of the flowers to insects in insect-pollinated plants. This strategy is not without risks, as the flowers can be damaged by frost or, in dry season regions, result in water stress on the plant. Nevertheless, there is much less branch and trunk breakage from glaze ice storms when leafless, and plants can reduce water loss due to the reduction in availability of liquid water during cold winter days. Leaf drop or abscission involves complex physiological signals and changes within plants. The process of photosynthesis steadily degrades the supply of chlorophylls in foliage; plants normally replenish chlorophylls during the summer months. When autumn arrives and the days are shorter or when plants are drought-stressed, deciduous trees decrease chlorophyll pigment production, allowing other pigments present in the leaf to become apparent, resulting in non-green coloured foliage.

The brightest leaf colours are produced when days grow short and nights are cool, but remain above freezing. These other pigments include carotenoids that are yellow, brown, and orange. Anthocyanin pigments produce red and purple colours, though they are not always present in the leaves. Rather, they are produced in the foliage in late summer, when sugars are trapped in the leaves after the process of abscission begins. Parts of the world that have showy displays of bright autumn colours are limited to locations where days become short and nights are cool. In other parts of the world, the leaves of deciduous trees simply fall off without turning the bright colours produced from the accumulation of anthocyanin pigments.

The beginnings of leaf drop starts when an abscission layer is formed between the leaf petiole and the stem. This layer is formed in the spring during active new growth of the leaf; it consists of layers of cells that can separate from each other. The cells are sensitive to a plant hormone called auxin that is produced by the leaf and other parts of the plant. When auxin coming from the leaf is produced at a rate consistent with that from the body of the plant, the cells of the abscission layer remain connected; in autumn, or when under stress, the auxin flow from the leaf decreases or stops, triggering cellular elongation within the abscission layer. The elongation of these cells break the connection between the different cell layers, allowing the leaf to break away from the plant. It also forms a layer that seals the break, so the plant does not lose sap. A number of deciduous plants remove nitrogen and carbon from the foliage before they are shed and store them in the form of proteins in the vacuoles of parenchyma cells in the roots and the inner bark. In the spring, these proteins are used as a nitrogen source during the growth of new leaves or flowers.

Function

Plants with deciduous foliage have both advantages and disadvantages compared to plants with evergreen foliage. Since deciduous plants lose their leaves to conserve water or to better survive winter weather conditions, they must regrow new foliage during the next suitable growing season; this uses resources which evergreens do not need to expend. Evergreens suffer greater water loss during the winter and they also can experience greater predation pressure, especially when small. Losing leaves in winter may reduce damage from insects; repairing leaves and keeping them functional may be more costly than just losing and regrowing them. Removing leaves also reduces cavitation which can damage xylem vessels in plants. This then allows deciduous plants to have xylem vessels with larger diameters and therefore a greater rate of transpiration (and hence CO2 uptake as this occurs when stomata are open) during the summer growth period.

Deciduous Woody Plants

The deciduous characteristic has developed repeatedly among woody plants. Trees include Maple, many Oaks, Elm, Aspen, and Birch, among others, as well as a number of coniferous genera, such as Larch and Metasequoia. Deciduous shrubs include honeysuckle, viburnum, and many others. Most temperate woody vines are also deciduous, including grapes, poison ivy, virginia creeper, wisteria, etc. The characteristic is useful in plant identification; for instance in parts of Southern California and the American Southeast, deciduous and evergreen oak species may grow side by side.

Periods of leaf fall often coincide with seasons: winter in the case of cool-climate plants or the dry-season in the case of tropical plants, however there are no deciduous species among tree-like monocotyledonous plants, e.g. palms, yuccas, and Dracaenas.

Regions

Forests where a majority of the trees lose their foliage at the end of the typical growing season are called deciduous forests. These forests are found in many areas worldwide and have distinctive ecosystems, understory growth, and soil dynamics.

Two distinctive types of deciduous forest are found growing around the world.

Temperate deciduous forest biomes are plant communities distributed in North and South America, Asia, Southern slopes of the Himalayas, Europe and for cultivation purposes in Oceania. They have formed under climatic conditions which have great seasonable temperature variability with growth occurring during warm summers and leaf drop in autumn and dormancy during cold winters. These seasonally distinctive communities have diverse life forms that are impacted greatly by the seasonality of their climate, mainly temperature and precipitation rates. These varying and regionally different ecological conditions produce distinctive forest plant communities in different regions.

Figure: *Dry-season deciduous tropical forest*

Tropical and subtropical deciduous forest biomes have developed in response not to seasonal temperature variations but to seasonal rainfall patterns. During prolonged dry periods the foliage is dropped to conserve water and prevent death from drought. Leaf drop is not seasonally dependent as it is in temperate climates, and can occur any time of year and varies by region of the world. Even within a small local area there can be variations in the timing and duration of leaf drop; different sides of the same mountain and areas that have high water tables or areas along streams and rivers can produce a patchwork of leafy and leafless trees.

leaf Colours

While you were playing in the hot sun during summer vacation the trees on the streets, in the parks, and in the forests were working hard to keep you cool. To feed the shiny green leaves that make shade, trees use sunlight to convert water and carbon dioxide into sugar. This is called *photosynthesis*.

Now it's autumn, and you're ready — okay, *almost* ready — to be back in school. Those hardworking trees, on the other hand, need to take a break from all that photosynthesizing. When leaves change colour from green to yellow, bright orange, or red, you'll know that trees are beginning their long winter's rest.

Where do Leaf Colours Come from?

Leaf colour comes from *pigments.* Pigments are natural substances produced by leaf cells. The three pigments that colour leaves are:

- chlorophyll (green)
- carotenoid (yellow, orange, and brown)
- anthocyanin (red)

Chlorophyll is the most important of the three. Without the chlorophyll in leaves, trees wouldn't be able to use sunlight to produce food.

Carotenoids create bright yellows and oranges in familiar fruits and vegetables. Corn, carrots, and bananas are just a few of the many plants coloured by carotenoid.

Anthocyanins add the colour red to plants, including cranberries, red apples, cherries, strawberries and others.

Chlorophyll and carotenoid are in leaf cells all the time during the growing season. But the chlorophyll covers the carotenoid — that's why summer leaves are green, not yellow or orange. Most anthocyanins are produced only in autumn, and only under certain conditions. Not all trees can make anthocyanin.

How do Leaves Change Colour?

As the Earth makes its 365-day journey around the sun, some parts of the planet will get fewer hours of sunlight at certain times of the year. In those regions, the days become shorter and the nights get longer. The temperature slowly drops. Autumn comes, and then winter.

Trees respond to the decreasing amount of sunlight by producing less and less chlorophyll. Eventually, a tree stops producing chlorophyll. When that happens, the carotenoid already in the leaves can finally show through. The leaves become a bright rainbow of glowing yellows, sparkling oranges and warm browns. What about red leaves? Read on.

Do Leaves Change because of Weather?

Perhaps you've noticed that in some years, the red fall colours seem brighter and more spectacular than in other years. The *temperature* and *cloud cover* can make a big difference in a tree's red colours from year to year. When a number of warm, sunny autumn days and cool but not freezing nights come one after the other, it's going to be a good year for reds. In the daytime, the leaves can produce lots of sugar, but the cool night temperatures prevent the sugar sap from flowing through the leaf veins and down into the branches and trunk. Anthocyanins to the rescue! Researchers have found out that anthocyanins are produced as a form of protection. They allow the plant to recover nutrients in the leaves before they fall off. This helps make sure that the tree will be ready for the next growing season. Anthocyanins give leaves their bright, brilliant shades of red, purple and crimson.

William Schumacher
Age 10

The yellow, gold and orange colours created by carotenoid remain fairly constant from year to year. That's because carotenoids are always present in leaves and the amount does not change in response to weather.

The amount of rain in a year also affects autumn leaf colour. A severe drought can delay the arrival of fall colours by a few weeks. A warm, wet period during fall will lower the intensity, or brightness, of autumn colours. A severe frost will kill the leaves, turning them brown and causing them to drop early. The best autumn colours come when there's been:

- a warm, wet spring
- a summer that's not too hot or dry, and
- a fall with plenty of warm sunny days and cool nights.

Can You Tell a Tree from its Colours?

You can use fall leaf colour to help identify different tree species. Look for these leaf colours on the trees in your neighbourhood:

- Oaks: red, brown or russet
- Hickories: golden bronze

- Dogwood: purple-red
- Birch: bright yellow
- paper birch
- yellow birch
- Poplar: golden yellow
- Maple trees show a whole range of colours:
- Sugar Maple: orange-red
- Black Maple: glowing yellow
- Red Maple: bright scarlet

Why Do Leaves Change Colour?

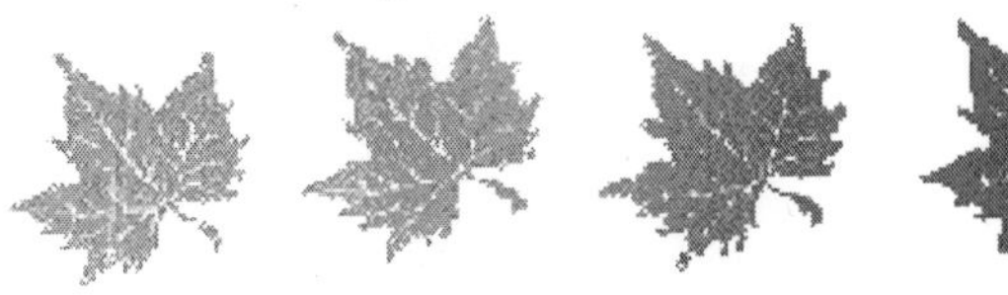

- Where do leaf colours come from?
- How do leaves change colour?
- Do leaves change colour because of weather?
- Why do leaves fall?
- 2013 Fall Colour Report (Dept. of Tourism)

While you were playing in the hot sun during summer vacation the trees on the streets, in the parks, and in the forests were working hard to keep you cool. To feed the shiny green leaves that make shade, trees use sunlight to convert water and carbon dioxide into sugar. This is called *photosynthesis*. Now it's autumn, and you're ready — okay, *almost* ready — to be back in school. Those hardworking trees, on the other hand, need to take a break from all that photosynthesizing. When leaves change colour from green to yellow, bright orange, or red, you'll know that trees are beginning their long winter's rest.

Where do leaf colours come from?

Leaf colour comes from *pigments*. Pigments are natural substances produced by leaf cells. The three pigments that colour leaves are:

- chlorophyll (green)
- carotenoid (yellow, orange, and brown)
- anthocyanin (red)

Chlorophyll is the most important of the three. Without the chlorophyll in leaves, trees wouldn't be able to use sunlight to produce food.*Carotenoids* create bright yellows and oranges in familiar fruits and vegetables. Corn, carrots, and bananas are just a few of the many plants coloured by carotenoid. *Anthocyanins* add the colour red to plants, including cranberries, red apples, cherries, strawberries and others. Chlorophyll and carotenoid are in leaf cells all the time during the growing season. But the chlorophyll covers the carotenoid — that's why summer leaves are green, not yellow or orange. Most anthocyanins are produced only in autumn, and only under certain conditions. Not all trees can make anthocyanin.

How do Leaves Change Colour?

As the Earth makes its 365-day journey around the sun, some parts of the planet will get fewer hours of sunlight at certain times of the year. In those regions, the days become shorter and the nights get longer. The temperature slowly drops. Autumn comes, and then winter. Trees respond to the decreasing amount of sunlight by producing less and less chlorophyll. Eventually, a tree stops producing chlorophyll. When that happens, the carotenoid already in the leaves can finally show through. The leaves become a bright rainbow of glowing yellows, sparkling oranges and warm browns. What about red leaves? Read on.

Do Leaves Change because of Weather?

Perhaps you've noticed that in some years, the red fall colours seem brighter and more spectacular than in other years. The *temperature* and *cloud cover* can make a big difference in a tree's red colours from year to year. When a number of warm, sunny autumn days and cool but not freezing nights come one after the other, it's going to be a good year for reds. In the daytime, the leaves can produce lots of sugar, but the cool night temperatures prevent the sugar sap from flowing through the leaf veins and down into the branches and trunk. Anthocyanins to the rescue! Researchers have found out that anthocyanins are produced as a form of protection. They allow the plant to recover nutrients in the leaves before they fall off. This helps make sure that the tree will be ready for the next growing season. Anthocyanins give leaves their bright, brilliant shades of red, purple and crimson. The yellow, gold and orange colours created by carotenoid remain fairly constant from year to year. That's because carotenoids are always present in leaves and the amount does not change in response to weather. The amount of rain in a year also affects autumn leaf colour. A severe drought can delay the arrival of fall colours by a few weeks. A warm, wet period during fall will lower the intensity, or brightness, of autumn colours. A severe frost will kill the leaves, turning them brown and causing them to drop early. The best autumn colours come when there's been:

- a warm, wet spring
- a summer that's not too hot or dry, and
- a fall with plenty of warm sunny days and cool nights.

Can You Tell a Tree from its Colours?

You can use fall leaf colour to help identify different tree species. Look for these leaf colours on the trees in your neighbourhood:

- Oaks: red, brown or russet
- Hickories: golden bronze
- Dogwood: purple-red
- Birch: bright yellow
- paper birch
- yellow birch
- Poplar: golden yellow
- Maple trees show a whole range of colours:
- Sugar Maple: orange-red
- Black Maple: glowing yellow
- Red Maple: bright scarlet

Why do Leaves Fall?

A tree's roots, branches and twigs can endure freezing temperatures, but most leaves are not so tough. On a broadleaf tree — say a maple or a birch — the tender thin leaves, made up of cells filled with water sap, will freeze in winter. Any plant tissue unable to live through the winter must be sealed off and shed to ensure the tree's survival. As sunlight decreases in autumn, the veins that carry sap into and out of a leaf gradually close. A layer of cells, called the *separation layer*, forms at the base of the leaf stem. When this layer is complete, the leaf is separated from the tissue that connected it to the branch, and it falls. Oak leaves are the exception.

The separation layer never fully detaches the dead oak leaves, and they remain on the tree through winter. Evergreen trees — pines, spruces, cedars and firs — don't lose their leaves, or needles, in winter. The needles are covered with a heavy wax coating and the fluids inside the cells contain substances that resist freezing. Evergreen leaves can live for several years before they fall and are replaced by new growth. On the ground, fallen leaves are broken down by bacteria, fungi, earthworms and other organisms. The decomposed leaves restock the soil with nutrients, and become part of the spongy humus layer on the forest floor that absorbs and holds rainfall. In nature, nothing goes to waste!

The Chemistry of Autumn Colours

Every autumn across the Northern Hemisphere, diminishing daylight hours and falling temperatures induce trees to prepare for winter. In these preparations, they shed billions of tons of leaves. In certain regions, such as our own, the shedding of leaves is preceded by a spectacular colour show. Formerly green leaves turn to brilliant shades of yellow, orange, and red. These colour changes are the result of transformations in leaf pigments.

The green pigment in leaves is chlorophyll. Chlorophyll absorbs red and blue light from the sunlight that falls on leaves. Therefore, the light reflected by the leaves is diminished in red and blue and appears green. The molecules of chlorophyll are large ($C_{55}H_{70}MgN_4O_6$). They are not soluble in the aqueous solution that fills plant cells. Instead, they are attached to the membranes of disc-like structures, called chloroplasts, inside the cells. Chloroplasts are the site of photosynthesis, the process in which light energy is converted to chemical energy. In chloroplasts, the light absorbed by chlorophyll supplies the energy used by plants to transform carbon dioxide and water into oxygen and carbohydrates, which have a general formula of $C_x(H_2O)_y$.

$$\text{x}\, CO_2 + \text{y}\, H_2O \xrightarrow[\text{chlorophyll}]{\text{light}} \text{x}\, O_2 + Cx(H_2O)y$$

In this endothermic transformation, the energy of the light absorbed by chlorophyll is converted into chemical energy stored in carbohydrates (sugars and starches). This chemical energy drives the biochemical reactions that cause plants to grow, flower, and produce seed.

Chlorophyll is not a very stable compound; bright sunlight causes it to decompose. To maintain the amount of chlorophyll in their leaves, plants continuously synthesize it. The synthesis of chlorophyll in plants requires sunlight and warm temperatures. Therefore, during summer chlorophyll is continuously broken down and regenerated in the leaves of trees.

Another pigment found in the leaves of many plants is carotene. Carotene absorbs blue-green and blue light. The light reflected from carotene appears yellow. Carotene is also a large molecule ($C_{40}H_{36}$) contained in the chloroplasts of many plants. When carotene and chlorophyll occur in the same leaf, together they remove red, blue-green, and blue light from sunlight that falls on the leaf. The light reflected by the leaf appears green. Carotene functions as an accessory absorber. The energy of the light absorbed by carotene is transferred to chlorophyll, which uses the energy in photosynthesis. Carotene is a much more stable compound than chlorophyll. Carotene persists in leaves even when chlorophyll has disappeared. When chlorophyll disappears from a leaf, the remaining carotene causes the leaf to appear yellow.

A third pigment, or class of pigments, that occur in leaves are the anthocyanins. Anthocyanins absorb blue, blue-green, and green light. Therefore, the light reflected by leaves containing anthocyanins appears red. Unlike chlorophyll and carotene, anthocyanins are not attached to cell membranes, but are dissolved in the cell sap. The colour produced by these pigments is sensitive to the pH of the cell sap. If the sap is quite acidic, the pigments impart a bright red colour; if the sap is less acidic, its colour is more purple. Anthocyanin pigments are responsible for the red skin of ripe apples and the purple of ripe grapes. Anthocyanins are formed by a reaction between sugars and certain proteins in cell sap. This reaction does not occur until the concentration of sugar in the sap is quite high. The reaction also requires light. This is why apples often appear red on one side and green on the other; the red side was in the sun and the green side was in shade.

During summer, the leaves of trees are factories producing sugar from carbon dioxide and water by the action of light on chlorophyll.

Chlorophyll causes the leaves to appear green. (The leaves of some trees, such as birches and cottonwoods, also contain carotene; these leaves appear brighter green, because carotene absorbs blue-green light.) Water and nutrients flow from the roots, through the branches, and into the leaves.

The sugars produced by photosynthesis flow from the leaves to other parts of the tree, where some of the chemical energy is used for growth and some is stored. The shortening days and cool nights of autumn trigger changes in the tree. One of these changes is the growth of a corky membrane between the branch and the leaf stem. This membrane interferes with the flow of nutrients into the leaf. Because the nutrient flow is interrupted, the production of chlorophyll in the leaf declines, and the green colour of the leaf fades. If the leaf contains carotene, as do the leaves of birch and hickory, it will change from green to bright yellow as the chlorophyll disappears. In some trees, as the concentration of sugar in the leaf increases, the sugar reacts to form anthocyanins. These pigments cause the yellowing leaves to turn red. Red maples, red oaks, and sumac produce anthocyanins in abundance and display the brightest reds and purples in the autumn landscape.

Figure: *Sugar Maple*

The range and intensity of autumn colours is greatly influenced by the weather. Low temperatures destroy chlorophyll, and if they stay above freezing, promote the formation of anthocyanins. Bright sunshine also destroys chlorophyll and enhances anthocyanin production. Dry weather, by increasing sugar concentration in sap, also increases the amount of anthocyanin. So the brightest autumn colours are produced when dry, sunny days are followed by cool, dry nights.

The Fascinatinating Story of How Leaves Change Colour

My wife and I have considered fall our favourite season for 31 years, whether the colours are brilliant and mesmerizing or muted and dull. We've tried repeatedly to find the reason, but can't quite pin one down. Probably because there are so many fond factors involved.

That we first met in the fall is probably the best reason (my wife made me write that). But there's something else, too. Perhaps it's the fragrance of smoke wafting from chimneys and mixing with the heady scent of the last cutting, the harvest, and fresh, chilly air.

Or maybe it's the promise of snows to come (another family favourite) and the coziness of a down comforter on a cold, autumn night. Or maybe it's the awe-inspiring, majestic colours. Fall foliage colour changes, though always the same process year after year, do have different results. The leaves can be varied and gorgeous, deeply coloured speactacles that take your breath away.

Looking at all of those features makes me think it's all of the above and that the absence of any one would deprive us of the awesomeness we've come to expect.

Regardless, it never occurred to either of us to question what motivates trees and their leaves to form a strong synergy during the spring and a "leaf-be-gone" strategy in the fall.

Through the process of photosynthesis, the leaves, at least in part, helped to make the tree's food, only to have trees deprive the leaves of chlorophyll (what makes them green and helps the tree produce glucose), force them to produce stark, luscious Fall Foliage colours, break away and become the next layer of soil on the forest floor.

As we know, the leaf colour change is a phenomenon unique to deciduous forests, a healthy mix of maple, birch, poplar, redbud, elm, sugar maple, evergreens, sumac, oak, dogwood, hickory, sweet gum, black gum, sourwood and ash trees.

Which Tree Leaves Turn What Colour?

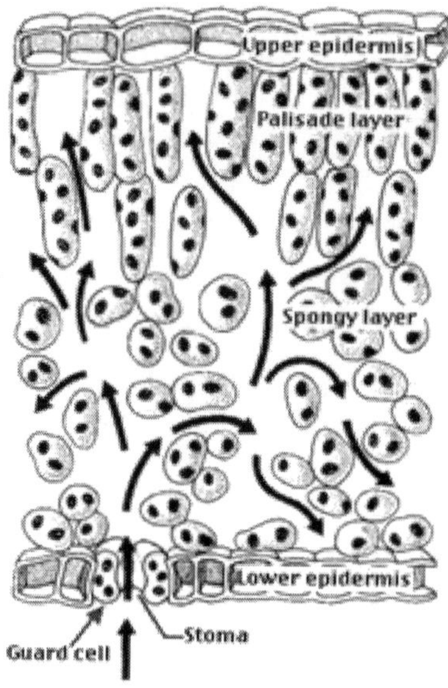

Leaf colour changes in New England fall foliage colour changes are determined by the tree's species. For instance, the evergreens, spruce, fir trees and rhododendrons and laurel, among others, are called evergreens for a reason. Even though some species will change the colour of their leaves, what remains at season's end stays green.

Birch, elm, poplar, redbud and hickory trees give us the gold and yellow hues that seem to dominate as the backdrop for the reds in autumn's colour scheme.

Red oak, hickory, ash, sugar maple, sweet gum, black gum, sourwood, red maple, as well as dogwoods give us the stunning red, deep magenta and purple colours, although they can produce yellow leaves. We get maroon from sumac.

Why and How do Leaves Change Colour?

Knowing which tree produces which colour is one thing. The big and growing question over the years has been why and how they change? The most popular belief in scientific circles holds that as there are more summer days behind than there are ahead, and as the days grow shorter and the nights longer and cooler, the leaf colour change is triggered within the trees.

The process begins with the severance of chlorophyll to the leaves and clothing them in their inherited colours before they fall off and glide to nature's woodland carpets.

As with all things scientific (or botany-related in this case), theories conceived by the specialists exist as slightly different spins on the phenomenon that visits New England and other parts of the country each year.

The Fall Foliage Colour Change Process

Without plants, trees in particular, I wouldn't be keyboarding this story now. They are nature's carbon dioxide scrubbers and oxygen providers. As we inhale, we're breathing IN oxygen produced by plants and trees. Our blood is oxygenated in our lungs and it then carries the oxygen to our tissues. What gets exhaled in that process is carbon dioxide.

If that builds up to certain levels, it can be fatal in humans, so trees feed us the life-nurturing gas we need to breath and protect us from harmful gases that would imperil our lives.

Water is taken from the soil by the tree's roots, and carbon dioxide from the air by the leaves. Sunlight then converts both water and carbon dioxide into glucose, a natural sugar, which becomes food for trees.

The catalyst for all this is chlorophyll. Under the green in the leaves of trees that produce yellow leaf colour lays xanthophyll, one of the yellow pigments that gives the leaves their yellow, gold and orange colours.

Some scientists conclude that the whole leaf isn't completely yellow under the green, but rather exists in small patches that spread in the fall. Others maintain that what we see in autumn is just as prevalent in July, but lies hidden under summer's green tints.

During the summer, some of the tree's food (glucose), is stored for winter reserves. When summer ebbs, the trees are able to decrease or cease food production and stop the flow of chlorophyll to the leaves. That eliminates the green colour.

In contrast, the leaves that give us the deep red and magenta colours appear only in the fall when the temperature ranges between freezing and 45 degrees and the trees have plenty of sun.

They aren't "under" the green in the leaves as is the case with the yellow species. Instead, some of the glucose produced when the tree manufactured its food is trapped in the leaves and when the chlorophyll disappears, the glucose reddens because it contains anthrocyanin, a red pigment. It's the same tincture that makes roses and geraniums red.

Finally, oak leaves contain waste from the tree, so they turn brown before falling to the forest floor.

While leaf colour is determined by tree type, colour saturation, or how bright the hues become, is affected by how cool and sunny fall days are. The cooler and sunnier, the brighter the colours. But summer weather can play a role as well. If there's too much or too little rain, for instance, the colours may not be as bright during the peak period.

What Role does Weather Play in Colour Saturation?

The biggest determinants of how deep the hues will become are water and air temperature.

Frosty nights that stay above freezing, and sunny, warm fall days are responsible for colour depth. Leaves turn much brighter under those conditions.

In the case of trees that yield yellow leaves, ever present are carotenoids and carotene. These form the yellow tints that lie beneath the green and emerge as all yellow, orange or gold in the fall. In the case of red leaves, this ideal weather produces anthocyanin tinctures. They produce radiant reds, magenta and sometimes purple leaves.

How moist the soil is as it relates to the temperatures can produce a spectacular colour array or one that's dull and flat. Ideally, if the preceding spring produced a good deal of rain and summer weather was "normal," neither too hot, nor too cold, not too wet, nor too dry, the best autumn colours are produced.

A summer drought will likely delay peak foliage. Too much summer rain can have a similar effect. In other words, if everything is not in optimum balance, the colours can be delayed and when the leaves change colour, they appear muted.

What Makes Leaves Fall?

In early autumn, in response to the shortening days and declining intensity of sunlight, trees begin the processes leading up to their baldness (no PC emails; I'm follicly challenged myself). The veins that carry fluids into and out of the leaf gradually close off as a layer of cells forms at the base of each leaf. These clogged veins trap sugars in the leaf and promote production of anthocyanins. Once this separation layer is complete and the connecting tissues are sealed off, the leaf is ready to fall.

As autumn begins to yield to winter, the colours fade and the stems binding leaf to tree branch begin to weaken. Before long, the leaves are literally hanging by threads—the tiny arteries that brought them water and glucose during the summer. Fall's winds, leaf pickers, wild animals, almost anything that brushes up against them snap them off the tree.

Fortunately, the colours remain after the leaves have dropped, but only for a day or so, so if you want to collect and press them, you can take them off the trees yourself, or if you see one you really like on the ground, you can collect that and press it, but do it quickly. Once they dry out, they get very fragile and break easily.

Eventually, all of the leaves are rendered brown by the presence of tannin; it's the same substance that gives tea leaves their brown colour after they're harvested.

We have very carefully selected and picked from red-leaf-coloured trees the highest quality red leaves we can find and my wife has mastered the art of folding them into what appears to be a rose.

How Does the Weather Affect the Colour of Fall Leaves?

As the weather turns cooler, the trees will once again put on their show of brightly coloured leaves. Many weather factors play into how intense the colours will be and how long the vivid leaves will be around.

After the leaves are fully developed on trees, they begin making and storing the carbohydrates that will be needed for the new tree growth in the following year. As the season progresses, in late summer or early fall, the trees enter a growth process that produces the colourful fall leaves.

As the tree grows throughout the spring and summer months, chlorophyll is constantly replaced in the leaves. The chlorophyll gives the leaves their green colour. As the nights get longer in the early fall, the cells near the juncture of the leaf and stem divide rapidly but do not expand. This action of the cells form a layer called the abscission layer. The abscission layer then blocks the transportation of materials from the leaf to the branch and from the roots to the leaves. As the chlorophyll is blocked from the leaves, it disappears completely from them. The lack of chlorophyll allows the yellow (xanthophylls) and orange (carotenoids) pigments to be visible. The red and purple pigments (anthocyanins) are manufactured from the sugars that are trapped in the leaf. These pigments in leaves are responsible for the vivid colour changes in the fall.

Temperature, sunlight and soil moisture all play a role in how the leaves will look in the fall.

- Abundant sunlight and low temperatures after the abscission layer forms cause the chlorophyll to be destroyed more rapidly.
- Cool air (especially at night) with a lot of daytime sunshine promote the formation of more red and purple pigments.
- Freezing conditions destroy the leaf's ability to manufacture the red and purple pigments. Early frost will end the colourful foliage.
- Drought during the growing season can cause the abscission layer to form early and cause the leaves to drop before they change colour.
- The best weather for brilliant fall foliage is a growing season with ample moisture followed by a dry, cool and sunny autumn with warm days and cool but frostless nights.
- Heavy wind or rain can cause the leaves to fall before they fully develop colour.

Travel to these locations this fall to take in some of the most breathtaking displays of fall foliage.

Charlottesville, Va., has oak trees that turn vibrant shades of red in the fall.

Milford, Pa., boasts golden hues of trees along the Delaware River.

Camden, Maine ,has trees of all colours that can be seen mirrored in the waters.

Sevierville, Tenn., has the Great Smokey Mountains in the east that are painted with brilliant fall foliage.

Glenwood Springs, Colo., shows off the fall foliage and snow-capped mountains in early October.

If you love to see the beautiful colours of fall, be sure to visit a tree-filled area near you this fall.

3

Diffusing Gases

Gas Exchange

Gas exchange occurs as a result of respiration, when carbon dioxide is excreted and oxygen taken up, and photosynthesis, when oxygen is excreted and carbon dioxide is taken up.

The rate of gas exchange is affected by:

- the area available for diffusion
- the distance over which diffusion occurs
- the concentration gradient across the gas exchange surface
- the speed with which molecules diffuse through membranes.

Efficient gas exchange systems must:

- have a large surface area to volume ratio
- be thin
- have mechanisms for maintaining steep concentration gradients across themselves
- be permeable to gases.

Single-celled organisms are aquatic and their cell surface membrane has a sufficiently large surface area to volume ratio to act as an efficient gas exchange surface.

In larger organisms, permeable, thin, flat structures have all the properties of efficient gas exchange surfaces but need water to prevent their dehydration and give them mechanical support. Since the solubility of oxygen in water is low, organisms that obtain their oxygen from water can maintain only a low metabolic rate. In small and thin organisms, the distance from gas exchange surface to the inside of the organism is short enough for diffusion of gases to be efficient. Diffusion gradients are maintained because gases are continually used up or produced. In larger organisms, simple diffusion is not an efficient way of transporting gases between cells in the body and the gas exchange surface. In many animals a blood circulatory system carries gases to and from the gas exchange surface. The gas-carrying capacity of the blood is increased by respiratory pigments, such as haemoglobin.

Animals with an internal gas exchange surface ventilate it by passing fresh air or water through their respiratory system. Air usually flows in and out through the same pathway; being light this requires little muscular activity. Denser water is passed in a one-directional pathway over gills.

In terrestrial plants diffusion of gases through pores is sufficient to service the few living cells in its stem cortex and its thalloid leaves. In the roots, gas exchange is restricted to a small permeable area.

Gas Exchange in Plants

In order to carry on photosynthesis, green plants need a supply of carbon dioxide and a means of disposing of oxygen. In order to carry on cellular respiration, plant cells need oxygen and a means of disposing of carbon dioxide (just as animal cells do).

Unlike animals, plants have no specialised organs for gas exchange (with the few inevitable exceptions!). The are several reasons they can get along without them:

- Each part of the plant takes care of its own gas exchange needs. Although plants have an elaborate liquid transport system, it does not participate in gas transport.
- Roots, stems, and leaves respire at rates much lower than are characteristic of animals. Only during photosynthesis are large volumes of gases exchanged, and each leaf is well adapted to take care of its own needs.
- The distance that gases must diffuse in even a large plant is not great. Each living cell in the plant is located close to the surface. While obvious for leaves, it is also true for stems. The only living cells in the stem are organised in thin layers just beneath the bark. The cells in the interior are dead and serve only to provide mechanical support.
- Most of the living cells in a plant have at least part of their surface exposed to air. The loose packing of parenchyma cells in leaves, stems, and roots provides an interconnecting system

of air spaces. Gases diffuse through air several thousand times faster than through water. Once oxygen and carbon dioxide reach the network of intercellular air spaces (arrows), they diffuse rapidly through them.

- Oxygen and carbon dioxide also pass through the cell wall and plasma membrane of the cell by diffusion. The diffusion of carbon dioxide may be aided by aquaporin channels inserted in the plasma membrane.

Leaves

The exchange of oxygen and carbon dioxide in the leaf (as well as the loss of water vapor in transpiration) occurs through pores called stomata (singular = stoma).

Normally stomata open when the light strikes the leaf in the morning and close during the night.

The immediate cause is a change in the turgor of the guard cells. The inner wall of each guard cell is thick and elastic. When turgor develops within the two guard cells flanking each stoma, the thin outer walls bulge out and force the inner walls into a crescent shape. This opens the stoma. When the guard cells lose turgor, the elastic inner walls regain their original shape and the stoma closes.

Time	*Osmotic Pressure, lb/in*
7 A.M.	212
11 A.M.	456
5 P.M.	272
12 midnight	191

The table shows the osmotic pressure measured at different times of day in typical guard cells. The osmotic pressure within the other cells of the lower epidermis remained constant at 150 lb/in (~1000 kilopascal, kPa). When the osmotic pressure of the guard cells became greater than that of the surrounding cells, the stomata opened. In the evening, when the osmotic pressure of the guard cells dropped to nearly that of the surrounding cells, the stomata closed.

Opening Stomata

The increase in osmotic pressure in the guard cells is caused by an uptake of potassium ions (K^+). The concentration of K^+ in open guard cells far exceeds that in the surrounding cells. This is how it accumulates:

- Blue light is absorbed by phototropin which activates
- a proton pump (an H^+-ATPase) in the plasma membrane of the guard cell.
- ATP, generated by the light reactions of photosynthesis, drives the pump.
- As protons (H^+) are pumped out of the cell, its interior becomes increasingly negative.
- This attracts additional potassium ions into the cell, raising its osmotic pressure.

Closing Stomata

Although open stomata are essential for photosynthesis, they also expose the plant to the risk of losing water through transpiration. Some 90% of the water taken up by a plant is lost in transpiration.

In angiosperms and gymnosperms (but not in ferns and lycopsids), Abscisic acid (ABA) is the hormone that triggers closing of the stomata when soil water is insufficient to keep up with transpiration (which often occurs around mid-day).

The mechanism:

- ABA binds to receptors at the surface of the plasma membrane of the guard cells.
- The receptors activate several interconnecting pathways which converge to produce
  - o a rise in pH in the cytosol;
  - o transfer of Ca^{2+} from the vacuole to the cytosol.
- These changes
  - o stimulate the loss of negatively-charged ions (anions), especially NO_3^- and Cl^-, from the cell and also
  - o the loss of K^+ from the cell.
- The loss of these solutes in the cytosol reduces the osmotic pressure of the cell and thus turgor.
- The stomata close.

Open stomata also provide an opening through which bacteria can invade the interior of the leaf. However, guard cells have receptors that can detect the presence of molecules associated with bacteria called pathogen-associated molecular patterns (PAMPs). LPS and flagellin are examples. When the guard cells detect these PAMPs, ABA mediates closure of the stoma and thus close the door to bacterial entry.

This system of innate immunity resembles that found in animals. Link to discussion.

Density of Stomata

The density of stomata produced on growing leaves varies with such factors as:

- the temperature, humidity, and light intensity around the plant;
- and also, as it turns out, the concentration of carbon dioxide in the air around the leaves. The relationship is inverse; that is, as the concentration of CO_2 goes up, the number of stomata produced goes down, and vice versa. Some evidence:
    - o Plants grown in an artificial atmosphere with a high level of CO_2 have fewer stomata than normal.
    - o Herbarium specimens reveal that the number of stomata in a given species has been declining over the last 200 years — the time of the industrial revolution and rising levels of CO_2 in the atmosphere.

These data can be quantified by determining the stomatal index: the ratio of the number of stomata in a given area divided by the total number of stomata and other epidermal cells in that same area.

How does the plant determine how many stomata to produce? It turns out that the mature leaves on the plant detect the conditions around them and send a signal (its nature still unknown) that adjusts the number of stomata that will form on the developing leaves.

Two experiments (reported by Lake et al., in Nature, 411:154, 10 May 2001):

- When the mature leaves of the plant (Arabidopsis) are encased in glass tubes filled with high levels (720 ppm) of CO_2, the developing leaves have fewer stomata than normal even though they are growing in normal air (360 ppm).
- Conversely, when the mature leaves are given normal air (360 ppm CO_2) while the shoot is exposed to high CO_2 (720 ppm), the new leaves develop with the normal stomatal index.

*One signal that increases stomatal density in 2-day-old Arabidopsis seedlings (a different experimental setup than the one above) is a 45-amino acid peptide called stomagen that is released by mesophyll cells and induces the formation of stomata in the epidermis above.

Stomata Reveal Past Carbon Dioxide Levels

Because CO_2 levels and stomatal index are inversely related, could fossil leaves tell us about past levels of CO_2 in the atmosphere? Yes. As reported by Gregory Retallack (in Nature, 411:287, 17 May 2001), his study of the fossil leaves of the ginkgo and its relatives shows:

- their stomatal indices were high:
    - late in the Permian period (275–290 million years ago) and again
    - in the Pleistocene epoch (1–8 million years ago).

    Both these periods are known from geological evidence to have been times of
    - low levels of atmospheric carbon dioxide and
    - ice ages (with glaciers).
- Conversely, stomatal indices were low during the Cretaceous period, a time of high CO_2 levels and warm climate.

These studies also lend support to the importance of carbon dioxide as a greenhouse gas playing an important role in global warming.

Roots and Stems

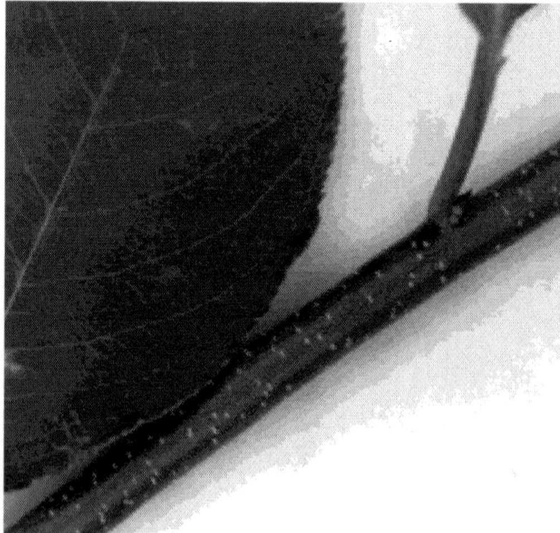

Woody stems and mature roots are sheathed in layers of dead cork cells impregnated with suberin — a waxy, waterproof (and airproof) substance. So cork is as impervious to oxygen and carbon dioxide as it is to water.

However, the cork of both mature roots and woody stems is perforated by nonsuberized pores called lenticels. These enable oxygen to reach the intercellular spaces of the interior tissues and carbon dioxide to be released to the atmosphere.

The photo shows the lenticels in the bark of a young stem.

In many annual plants, the stems are green and almost as important for photosynthesis as the leaves. These stems use stomata rather than lenticels for gas exchange.

Calculating Important Parameters in Leaf Gas Exchange

Evaporation (E) can be defined using Fick's law:

$$E = \frac{(e_i - e_a)}{\mathrm{P_r}}$$

Web Equation 9.4.1

where E is the evaporation rate; e_i and e_a are the partial pressures of water inside the leaf and in the ambient air, respectively; P is the atmospheric pressure; and r is the stomatal resistance. The ratio $(e_i - e_a)/P$ is the difference of water vapor between the inside and the outside of the leaf (in mole fraction).

The concept of a resistance to gas diffusion is intuitively obvious but has the disadvantage that evaporation is inversely related to r. However, one can define conductance to water vapor (g) as the inverse of r. In contrast to r, g is directly related to E, and it is the parameter usually used in gas exchange analysis. Web Equation 9.4.1 can thus be rearranged as follows:

$$E = \frac{(e_i - e_a)g}{p}$$

Web Equation 9.4.2

and Web Equation 9.4.2 can be used to solve for g:

$$g = \frac{EP}{(e_i - e_a)}$$

Web Equation 9.4.3

The rate of evaporation from a leaf can be determined with devices that measure the amount of water leaving the leaf, and the vapor pressure of water in the ambient air can also be measured. The air in the intercellular spaces of the leaf is assumed to be at 100% humidity, and the vapor pressure of water at 100% relative humidity is a function of temperature. Therefore, we determine e_i by measuring the leaf temperature. Since CO_2 diffuses along the same pathway as water, we can write an equation defining photosynthetic CO_2 assimilation (A) based on Web Equation 9.4.2:

$$A = \frac{(C_a - C_i)}{1.6\text{Pr}} = \frac{(Ca - Ci)g}{1.6P}$$

Web Equation 9.4.4

where C_a and C_i are the partial pressures of CO_2 in ambient air and in the intercellular spaces of the leaf, respectively. The CO_2 molecule is larger than H_2O and therefore has a smaller diffusion coefficient; the difference between the two diffusion coefficients has been empirically determined to be 1.6 (von Caemmerer and Farquhar 1981). Use of this correction factor in Web Equation 9.4.4 allows us to convert the measured resistance to water vapor diffusion into a resistance to CO_2 diffusion.

From Web Equation 9.4.4 we can calculate the partial pressure of CO_2 in the intercellular spaces of the leaf:

$$C_i = C_a - \frac{1.6AP}{g}$$

Web Equation 9.4.5

Since all the variables on the right-hand side of Web Equation 9.4.5 can be either measured or calculated, this equation is the basis for the calculation of C_i in experiments in which A is measured as a function of C_i (von Caemmerer and Farquhar 1981).

Water use efficiency can be quantified by a combination of Web Equations 9.4.2 and 9.4.4:

$$\frac{A}{E} = \frac{(C_a - C_i)}{1.6(e_i - e_a)}$$

Web Equation 9.4.6

Let's assume a leaf temperature of 30°C and an atmosphere with relative humidity of 50%. For a C_3 plant, $C_a - C_i$ is typically 10 Pa. At a leaf temperature of 30°C and a relative humidity of 50%, $e_i - e_a = 2.1 \times 10^3$ Pa. From Web Equation 9.4.6, we compute that A/E = 0.003 mol CO_2 (mol $H_2O)^{-1}$, or a transpiration ratio (E/A) of 336. If water availability is reduced, stomata close and evaporation decreases, leading to improved water use efficiency. Total CO_2 assimilation also decreases, but the plant conserves water and increases its chances of survival. These equations have been used very successfully to advance our understanding of photosynthesis and the responses of plants to their environments. However, every technique has limitations, and these equations assume uniform stomatal behaviour across the leaf surface. Problems might arise when some stomata are closed and others are open, especially in leaves in which discrete regions of mesophyll are sealed off from each other by leaf veins. Under some conditions, in one mesophyll region the stomata are open and there is a correspondingly high photosynthetic rate, whereas in an immediately adjacent region the stomata are partially closed and the mesophyll has a low photosynthetic rate.

This phenomenon is termed *patchy stomata* and might arise under low fluence rates of illumination, or when leaves are stressed (Terashima 1992). Although patchy stomata may reflect normal interactions between stomata and leaf mesophyll (Siebke and Weis 1995), patchiness can lead to inaccurate determinations of C_i. In other experimental systems, patchiness has been shown not to affect calculations of intercellular carbon dioxide by conventional gas exchange methods (Kueppers et al. 1999).

4

Leaking Water

Stoma

In botany, a stoma (plural *stomata*) (occasionally called a stomate, plural stomates) (from Greek óôüìá, "mouth") is a pore, found in the epidermis of leaves, stems and other organs that is used to control gas exchange. The pore is bordered by a pair of specialised parenchyma cells known as guard cells that are responsible for regulating the size of the opening.

The term is also used collectively to refer to an entire stomatal complex, both the pore itself and its accompanying guard cells. Air containing carbon dioxide and oxygen enters the plant through these openings and is used in photosynthesis in the mesophyll cells (parenchyma cells with chloroplasts) and respiration, respectively. Oxygen produced as a by-product of photosynthesis diffuses out to the atmosphere through these same openings. Also, water vapor is released into the atmosphere through these pores in a process called transpiration.

Stomata are present in the sporophyte generation of all land plant groups except liverworts. Dicotyledons usually have more stomata on the lower epidermis than the upper epidermis. Monocotyledons, on the other hand, usually have the same number of stomata on the two epidermes. In plants with floating leaves, stomata may be found only on the upper epidermis; submerged leaves may lack stomata entirely.

Function

CO_2 Gain and Water Loss

Carbon dioxide, a key reactant in photosynthesis, is present in the atmosphere at a concentration of about 390 ppm (as of December

2011). Most plants require the stomata to be open during daytime. The problem is that the air spaces in the leaf are saturated with water vapour, which exits the leaf through the stomata (this is known as transpiration). Therefore, plants cannot gain carbon dioxide without simultaneously losing water vapour.

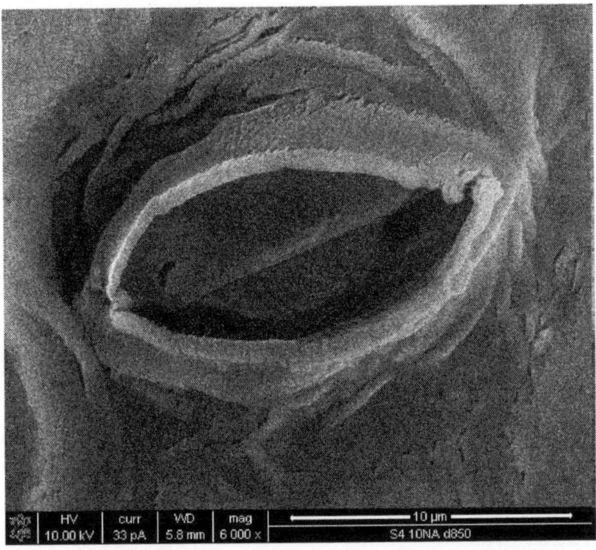

Figure: *This is an electron micrograph of a stoma from a Brassica chinensis (Bok Choy) leaf.*

Alternative Approaches

Ordinarily, carbon dioxide is fixed to ribulose-1,5-bisphosphate (RuBP) by the enzyme RuBisCO in mesophyll cells exposed directly to the air spaces inside the leaf. This exacerbates the transpiration problem for two reasons: first, RuBisCo has a relatively low affinity for carbon dioxide, and second, it fixes oxygen to RuBP, wasting energy and carbon in a process called photorespiration. For both of these reasons, RuBisCo needs high carbon dioxide concentrations, which means wide stomatal apertures and, as a consequence, high water loss.

Narrower stomatal apertures can be used in conjunction with an intermediary molecule with a high carbon dioxide affinity, PEPcase (Phosphoenolpyruvate carboxylase). Retrieving the products of carbon fixation from PEPCase is in an energy-intensive process, however. As a result, the PEPCase alternative is preferable only where water is limiting but light is plentiful, or where high temperatures increase the solubility of oxygen relative to that of carbon dioxide, magnifying RuBisCo's oxygenation problem.

CAM Plants

A group of mostly desert plants called "CAM" plants (Crassulacean acid metabolism, after the family Crassulaceae, which includes the species in which the CAM process was first discovered) open their stomata at night (when water evaporates more slowly from leaves for a given degree of stomatal opening), use PEPcarboxylase to fix carbon dioxide and store the products in large vacuoles. The following day, they close their stomata and release the carbon dioxide fixed the previous night into the presence of RuBisCO. This saturates RuBisCO with carbon dioxide, allowing minimal photorespiration. This approach, however, is severely limited by the capacity to store fixed carbon in the vacuoles, so it is preferable only when water is severely limiting.

Opening and Closure

However, most plants do not have the aforementioned facility and must therefore open and close their stomata during the daytime in response to changing conditions, such as light intensity, humidity, and carbon dioxide concentration. It is not entirely certain how these responses work. However, the basic mechanism involves regulation of osmotic pressure.

When conditions are conducive to stomatal opening (e.g., high light intensity and high humidity), a proton pump drives protons (H^+) from the guard cells. This means that the cells' electrical potential becomes increasingly negative. The negative potential opens potassium voltage-gated channels and so an uptake of potassium ions (K^+) occurs. To maintain this internal negative voltage so that entry of potassium ions does not stop, negative ions balance the influx of potassium. In some cases, chloride ions enter, while in other plants the organic ion malate is produced in guard cells. This increase in solute concentration lowers the water potential inside the cell, which results in the diffusion of water into the cell through osmosis. This increases the cell's volume and turgor pressure. Then, because of rings of cellulose microfibrils that prevent the width of the guard cells from swelling, and thus only allow the extra turgor pressure to elongate the guard cells, whose ends are held firmly in place by surrounding epidermal cells, the two guard cells lengthen by bowing apart from one another, creating an open pore through which gas can move.

When the roots begin to sense a water shortage in the soil, abscisic acid (ABA) is released. ABA binds to receptor proteins in the guard cells' plasma membrane and cytosol, which first raises the pH of the cytosol of the cells and cause the concentration of free Ca^{2+} to increase in the cytosol due to influx from outside the cell and release of Ca^{2+} from internal stores such as the endoplasmic reticulum and vacuoles. This causes the chloride (Cl^-) and inorganic ions to exit the cells. Second, this stops the uptake of any further K^+ into the cells and, subsequently, the loss of K^+. The loss of these solutes causes an increase in water potential, which results in the diffusion of water back out of the cell by osmosis. This makes the cell plasmolysed, which results in the closing of the stomatal pores.

Guard cells have more chloroplasts than the other epidermal cells from which guard cells are derived. Their function is controversial.

Inferring Stomatal Behaviour from Gas Exchange

The degree of stomatal resistance can be determined by measuring leaf gas exchange of a leaf. The transpiration rate is dependent on the diffusion resistance provided by the stomatal pores, and also on the humidity gradient between the leaf's internal air spaces and the outside air. Stomatal resistance (or its inverse, stomatal conductance) can therefore be calculated from the transpiration rate and humidity gradient. This allows scientists to investigate how stomata respond to changes in environmental conditions, such as light intensity and concentrations of gases such as water vapor, carbon dioxide, and ozone. Evaporation (*E*) can be calculated as;

$$E = (e_i - e_a) / Pr$$

where e_i and e_a are the partial pressures of water in the leaf and in the ambient air, respectively, *P* is atmospheric pressure, and *r* is stomatal resistance. The inverse of *r* is conductance to water vapor (*g*), so the equation can be rearranged to;

$$E = (e_i - e_a) g / P$$

and solved for *g*;

$$g = EP / (e_i - e_a)$$

Photosynthetic CO_2 assimilation (*A*) can be calculated from

$$A = (C_a - C_i) g / 1.6P$$

where C_a and C_i are the atmospheric and sub-stomatal partial pressures of CO_2, respectively. The rate of evaporation from a leaf can be determined using a photosynthesis system. These scientific instruments measure the amount of water vapour leaving the leaf and the vapor pressure of the ambient air. Photosynthetic systems may calculate water use efficiency (*A/E*), *g*, intrinsic water use efficiency (*A/g*), and C_i. These scientific instruments are commonly used by plant physiologists to measure CO_2 uptake and thus measure photosynthetic rate.

Stomata and Climate Change

Response of Stomata to Environmental Factors

Photosynthesis, plant water transport (xylem) and gas exchange are regulated by stomatal function which is important in the functioning of plants. Stomatal density and aperture (length of stomata) varies under a number of environmental factors such as atmospheric CO_2 concentration, light intensity, air temperature and photoperiod (daytime duration).

Decreasing stomatal density is one way plants have responded to the increase in concentration of atmospheric CO_2 ($[CO_2]_{atm}$). Although changes in $[CO_2]_{atm}$ response is the least understood mechanistically, this stomatal response has begun to plateau where it is soon expected to impact transpiration and photosynthesis processes in plants.

Future Adaptations during Climate Change

It is expected for $[CO_2]_{atm}$ to reach 500-1000 ppm by 2100. 96% of the past 400 000 years experienced below 280 ppm CO_2 levels. From this figure, it is highly probable that genotypes of today's plants diverged from their pre-industrial relative.

The gene *HIC* (high carbon dioxide) encodes a negative regulator for the development of stomata in plants. Research into the *HIC* gene using *Arabidopsis thaliana* found no increase of stomatal development in the dominant allele, but in the 'wild type' recessive allele showed a large increase, both in response to rising CO_2 levels in the atmosphere. These studies imply the plants response to changing CO_2 levels is largely controlled by genetics.

Agricultural Implications

In the face of ecological contingencies such as increasing temperatures, changes in rainfall patterns, long term climate change, and biotic influences of human management interventions, it is expected to reduce the production and quality of food and have a negative impact on agricultural production.

The CO_2 fertiliser effect has been greatly overestimated during Free-Air Carbon dioxide Enrichment (FACE) experiments where results show increased CO_2 levels in the atmosphere enhances photosynthesis, reduce transpiration, and increase water use efficiency (WUE). Increased biomass is one of the effects with simulations from experiments predicting a 5–20% increase in crop yields at 550 ppm of CO_2. Rates of leaf photosynthesis were shown to increase by 30-50% in C3 plants, and 10-25% in C4 under doubled CO_2 levels. The existence of a feedback mechanism results a phenotypic plasticity in response to $[CO_2]_{atm}$ that may have been an adaptive trait in the evolution of plant respiration and function.

Predicting how stomata performs during adaptation is useful for understanding the productivity of plant systems for both natural and agricultural systems. Plant breeders and farmers are beginning to work together using evolutionary and participatory plant breeding to find the best suited species such as heat and drought resistant crop varieties that could naturally evolve to the change in the face of food security challenges.

Evolution

Development: There are three major epidermal cell types which all ultimately derive from the L1 tissue layer of the shoot apical meristem, called protodermal cells: trichomes, pavement cells and guard cells, all of which are arranged in a non-random fashion.

Production is reliant on interactions between SPCH (speechless), EPF (downregulates stomata), TMM (too many mouths, downregulates stomata) and stomagen (upregulates stomata, inhibits SPCH), ERL and YODA downregulate stomata too.

Stomata positioning is down to CO2 activating EPF1, which activates TMM/ERL which together activate YODA, YODA in turn inhibits SPCH, in turn SPCH activation decreases, allowing asymmetry.

An asymmetrical cell division occurs in protodermal cells resulting in one large cell that is fated to become a pavement cell and a smaller cell called a meristemoid that will eventually differentiate into the guard cells that surround a stoma. This meristemoid then divides asymmetrically one to three times before differentiating into a guard mother cell. The guard mother cell then makes one symmetrical division, which forms a pair of guard cells.

Stomata as Pathogenic Pathways

Stomata are an obvious hole in the leaf by which, as was presumed for a while, pathogens can enter unchallenged. However, it has been recently shown that stomata do in fact sense the presence of some, if not all, pathogens. However, with the virulent bacteria applied to *Arabidopsis* plant leaves in the experiment, the bacteria released the chemical coronatine, which forced the stomata open again within a few hours.

Transpiration

This article is about plant transpiration.

1- Water is passively transported into the roots and then into the xylem.

Figure: *Overview of transpiration.*

2- The forces of cohesion and adhesion cause the water molecules to form a column in the xylem.

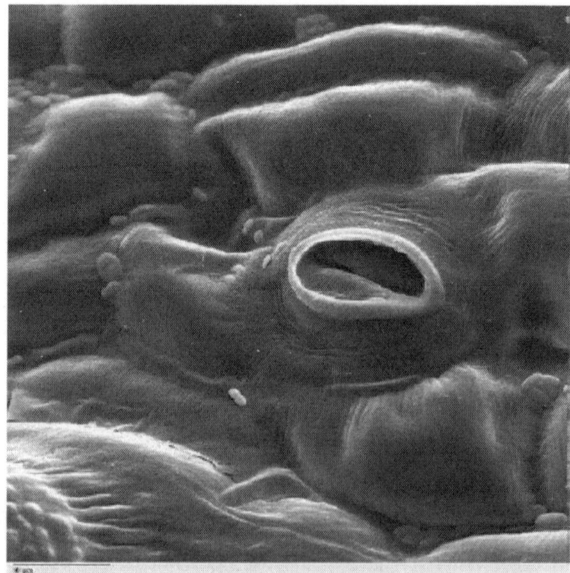

Figure: *Stoma in a tomato leaf shown via colourized scanning electron microscope*

3- Water moves from the xylem into the mesophyll cells, evaporates from their surfaces and leaves the plant by diffusion through the stomata.

Figure: *The clouds in this image of the Amazon Rainforest are a result of transpiration.*

Transpiration is the process of water movement through a plant and its evaporation from aerial parts, such as from leaves but also from stems and flowers. Leaf surfaces are dotted with pores which are called stomata, and in most plants they are more numerous on the undersides of the foliage. The stomata are bordered by guard cells and their stomatal accessory cells (together known as stomatal complex) that open and close the pore. Transpiration occurs through the stomatal apertures, and can be thought of as a necessary "cost" associated with the opening of the stomata to allow the diffusion of carbon dioxide gas from the air for photosynthesis. Transpiration also cools plants, changes osmotic pressure of cells, and enables mass flow of mineral nutrients and water from roots to shoots.

Mass flow of liquid water from the roots to the leaves is driven in part by capillary action, but primarily driven by water potential differences. In taller plants and trees, the force of gravity can only be overcome by the decrease in hydrostatic (water) pressure in the upper parts of the plants due to the diffusion of water out of stomata into the atmosphere. Water is absorbed at the roots by osmosis, and any dissolved mineral nutrients travel with it through the xylem.

Regulation

Plants regulate the rate of transpiration by the degree of stomatal opening. The rate of transpiration is also influenced by the evaporative demand of the atmosphere surrounding the leaf such as humidity, temperature, wind and incident sunlight. Soil water supply and soil temperature can influence stomatal opening, and thus transpiration rate.

The amount of water lost by a plant also depends on its size and the amount of water absorbed at the roots. Transpiration accounts for most of the water loss by a plant, but some direct evaporation also takes place through the cuticle of the leaves and young stems. Transpiration serves to evaporatively cool plants as the escaping water vapor carries away heat energy.

This table summarizes the factors that affect the rates of transpiration.

Feature	*How This Affects Transpiration*
Number of leaves	More leaves (or spines, or other photosynthesizing organs) means a bigger surface area and more stomata for gaseous exchange. This will result in greater water loss.
Number of stomata	More stomata will provide more pores for transpiration.
Size of the leaf	A leaf with a bigger surface area will transpire faster than a leaf with a smaller surface area.
Presence of plant cuticle	A waxy cuticle is relatively impermeable to water and water vapour and reduces evaporation from the plant surface except via the stomata. A reflective cuticle will reduce solar heating and temperature rise of the leaf, helping to reduce the rate of evaporation. Tiny hair-like structures called trichomes on the surface of leaves also can inhibit water loss by creating a high humidity environment at the surface of leaves. These are some examples of the adaptations of plants for conservation of water that may be found on many xerophytes.
Light supply	The rate of transpiration is controlled by stomatal aperture, and these small pores open especially for photosynthesis. While there are exceptions to this (such as night or "CAM photosynthesis"), in general a light supply will encourage open stomata.

Temperature	Temperature affects the rate in two ways: 1) An increased rate of evaporation due to a temperature rise will hasten the loss of water.2) Decreased relative humidity outside the leaf will increase the water potential gradient.
Relative humidity	Drier surroundings gives a steeper water potential gradient, and so increases the rate of transpiration.
Wind	In still air, water lost due to transpiration can accumulate in the form of vapor close to the leaf surface. This will reduce the rate of water loss, as the water potential gradient from inside to outside of the leaf is then slightly less. Wind blows away much of this water vapor near the leaf surface, making the potential gradient steeper and speeding up the diffusion of water molecules into the surrounding air. Even in wind, though, there is some accumulation of water vapor in a thin boundary layer of slower moving air next to the leaf surface. The stronger the wind, the thinner this layer, and the steeper the water potential gradient. Also, the bigger the leaf, the greater the average thickness of the boundary layer, which means a bigger leaf will have a slightly slower transpiration rate per unit area (although a higher transpiration rate overall).
Water supply	Water stress caused by restricted water supply from the soil may result in stomatal closure and reduce the rates of transpiration.

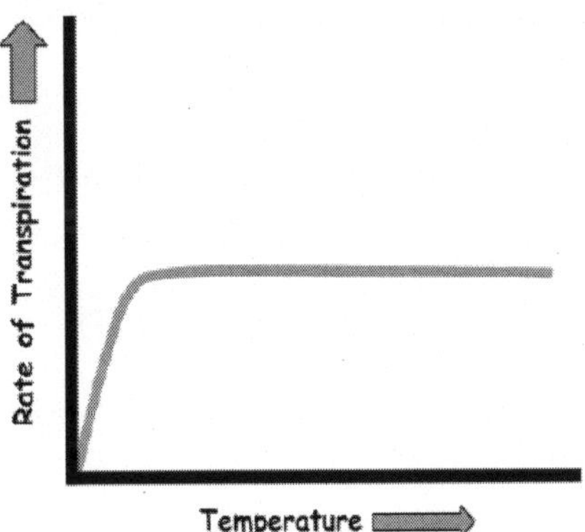

Figure: *The effect of temperature on the transpiration rate of plants.*

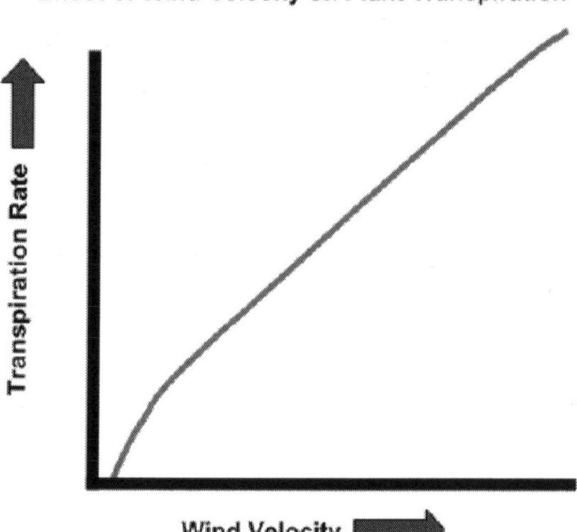

Figure: *The effect of wind velocity on the transpiration rate of plants.*

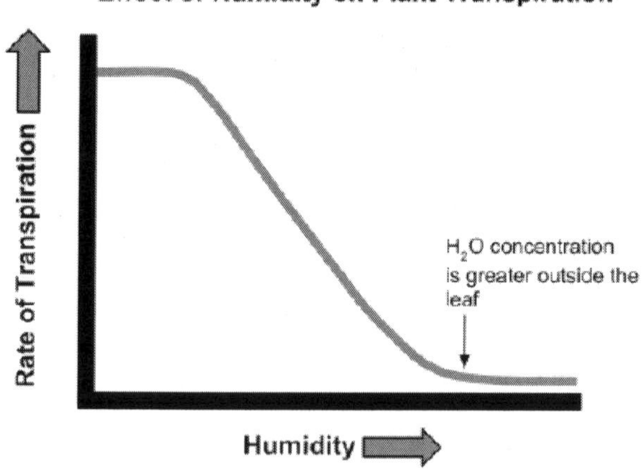

Figure: *The effect of humidity on the transpiration rate of plants.*

A fully grown tree may lose several hundred of gallons of water through its leaves on a hot, dry day. The transpiration ratio is the ratio of the mass of water transpired to the mass of dry matter produced; the transpiration ratio of crops tends to fall between 200 and 1000 (*i.e.*, crop plants transpire 200 to 1000 kg of water for every kg of dry matter produced).

Transpiration rates of plants can be measured by a number of techniques, including potometres, lysimetres, porometres, photosynthesis systems and heat balance sap flow gauges. Isotope measurements indicate transpiration is the larger component of evapotranspiration. Desert plants have specially adapted structures, such as thick cuticles, reduced leaf areas, sunken stomata and hairs to reduce transpiration and conserve water.

Many cacti conduct photosynthesis in succulent stems, rather than leaves, so the surface area of the shoot is very low. Many desert plants have a special type of photosynthesis, termed crassulacean acid metabolism or CAM photosynthesis, in which the stomata are closed during the day and open at night when transpiration will be lower.

Antitranspirant

Antitranspirants are compounds applied to the leaves of plants to reduce transpiration. They are used on Christmas trees, on cut flowers, on newly transplanted shrubs, and in other applications to preserve and protect plants from drying out too quickly. They have also been used to protect leaves from salt burn and fungal diseases.

Antitranspirants are of two types:

Metabolic Inhibitors

These reduce the stomatal opening and increase the leaf resistance to water vapour diffusion without affecting carbon dioxide uptake. Examples include phenylmercury acetate, absisic acid (ABA), and aspirin.

Film-Forming Antitranspirants

These form a colourless film on the leaf surface that allows diffusion of gases but not of water vapour. Examples include silicone oil and waxes.

Soil Plant Atmosphere Continuum

The Soil-Plant-Atmosphere Continuum (SPAC) is the pathway for water moving from soil through plants to the atmosphere.

The transport of water along this pathway occurs in components, variously defined among scientific disciplines:

- Soil physics characterizes water in soil in terms of tension,
- Physiology of plants and animals characterizes water in organisms in terms of *diffusion pressure deficit*, and
- Meteorology uses vapour pressure or relative humidity to characterize atmospheric water.

SPAC integrates these components and is defined as a:

...concept recognising that the field with all its components (soil, plant, animals and the ambient atmosphere taken together) constitutes a physically integrated, dynamic system in which the various flow processes involving energy and matter occur simultaneously and independently like links in the chain.

This characterises the state of water in different components of the SPAC as expressions of the energy level or water potential of each. Modelling of water transport between components relies on SPAC, as do studies of water potential gradients between segments.

Evapotranspiration

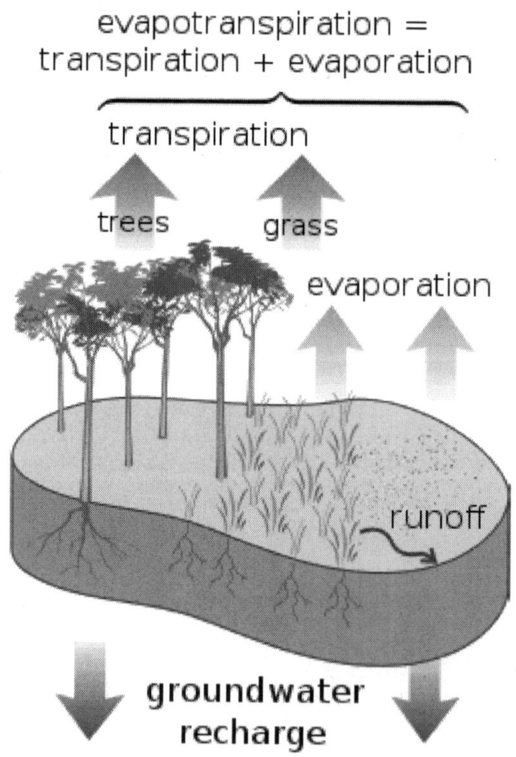

Figure: *Water cycle of the Earth's surface, showing the individual components of transpiration and evaporation that make up evapotranspiration. Other closely related processes shown are runoff and groundwater recharge.*

Evapotranspiration (ET) is the sum of evaporation and plant transpiration from the Earth's land and Ocean surface to the atmosphere. Evaporation accounts for the movement of water to the air from sources such as the soil, canopy interception, and waterbodies. Transpiration accounts for the movement of water within a plant and the subsequent loss of water as vapor through stomata in its leaves. Evapotranspiration is an important part of the water cycle. An element (such as a tree) that contributes to evapotranspiration can be called an evapotranspirator.

Potential evapotranspiration (PET) is a representation of the environmental demand for evapotranspiration and represents the evapotranspiration rate of a short green crop, completely shading the ground, of uniform height and with adequate water status in the soil profile. It is a reflection of the energy available to evaporate water, and of the wind available to transport the water vapour from the ground up into the lower atmosphere. Actual evapotranspiration is said to equal potential evapotranspiration when there is ample water.

Evapotranspiration and the Water Cycle

Evapotranspiration is a significant water loss from drainage basins. Types of vegetation and land use significantly affect evapotranspiration, and therefore the amount of water leaving a drainage basin. Because water transpired through leaves comes from the roots, plants with deep reaching roots can more constantly transpire water. Herbaceous plants generally transpire less than woody plants because they usually have less extensive foliage. Conifer forests tend to have higher rates of evapotranspiration than deciduous forests, particularly in the dormant and early spring seasons. This is primarily due to the enhanced amount of precipitation intercepted and evaporated by conifer foliage during these periods. Factors that affect evapotranspiration include the plant's growth stage or level of maturity, percentage of soil cover, solar radiation, humidity, temperature, and wind. Isotope measurements indicate transpiration is the larger component of evapotranspiration.

Through evapotranspiration, forests reduce water yield, except in unique ecosystems called cloud forests. Trees in cloud forests collect the liquid water in fog or low clouds onto their surface, which drips down to the ground. These trees still contribute to evapotranspiration, but often collect more water than they evaporate or transpire.

In areas that are not irrigated, actual evapotranspiration is usually no greater than precipitation, with some buffer in time depending on the soil's ability to hold water. It will usually be less because some water will be lost due to percolation or surface runoff. An exception is areas with high water tables, where capillary action can cause water from the groundwater to rise through the soil matrix to the surface. If potential evapotranspiration is greater than actual precipitation, then soil will dry out, unless irrigation is used.

Evapotranspiration can never be greater than PET, but can be lower if there is not enough water to be evaporated or plants are unable to transpire readily.

Estimating Evapotranspiration

Evapotranspiration can be measured or estimated using several methods.

Indirect Methods

Pan evaporation data can be used to estimate lake evaporation, but transpiration and evaporation of intercepted rain on vegetation are unknown. There are three general approaches to estimate evapotranspiration indirectly.

Catchment Water Balance

Evapotranspiration may be estimated by creating an equation of the water balance of a drainage basin. The equation balances the change in water stored within the basin (S) with inputs and exports:

$$\Delta S = P - ET - Q - D$$

The input is precipitation (P), and the exports are evapotranspiration (which is to be estimated), streamflow (Q), and groundwater recharge (D). If the change in storage, precipitation, streamflow, and groundwater recharge are all estimated, the missing flux, ET, can be estimated by rearranging the above equation as follows:

$$ET = P - \Delta S - Q - D$$

Hydrometeorological Equations

The most general and widely used equation for calculating reference ET is the Penman equation. The Penman-Monteith variation is recommended by the Food and Agriculture Organisation. The simpler Blaney-Criddle equation was popular in the Western United States for many years but it is not as accurate in regions with higher humidities. Other solutions used includes Makkink, which is simple but must be calibrated to a specific location, and Hargreaves. To convert the reference evapotranspiration to actual crop evapotranspiration, a crop coefficient and a stress coefficient must be used. Crop coefficients referred to in many hydrological models are themselves the result of equations that describe predictable variation in coefficient values depending upon plant conditions that change during periods for which the model is used. This is because crops are seasonal, perennial plants mature over multiple seasons, and stress responses can significantly depend upon many aspects of plant condition.

Energy Balance

A third methodology to estimate the actual evapotranspiration is the use of the energy balance.

$$\lambda E = R_n - G - H$$

where λE is the energy needed to change the phase of water from liquid to gas, R_n is the net radiation, G is the soil heat flux and H is the sensible heat flux. Using instruments like a scintillometre, soil heat flux plates or radiation metres, the components of the energy balance can be calculated and the energy available for actual evapotranspiration can be solved.

The SEBAL algorithm solves the energy balance at the earth surface using satellite imagery. This allows for both actual and potential evapotranspiration to be calculated on a pixel-by-pixel basis. Evapotranspiration is a key indicator for water management and irrigation performance. SEBAL can map these key indicators in time and space, for days, weeks or years.

Experimental Method for Measuring ET

One method for measuring ET is with a weighing lysimetre. The weight of a soil column is measured continuously and the change in storage of water in the soil is modelled by the change in weight. The change in weight is converted to units of length based on the surface area of the weighing lysimetre and the unit weight of water. ET is computed as the change in weight plus rainfall minus percolation.

Eddy Covariance

The most direct method of measuring evapotranspiration is with the eddy covariance technique in which fast fluctuations of vertical wind speed are correlated with fast fluctuations in atmospheric water vapor density. This directly estimates the transfer of water vapor (evapotranspiration) from the land (or canopy) surface to the atmosphere.

Evapotranspiration of Urban Landscape Plants

A review of ET measurement techniques for estimating the water requirements of urban landscape vegetation could be seen.

Potential Evapotranspiration

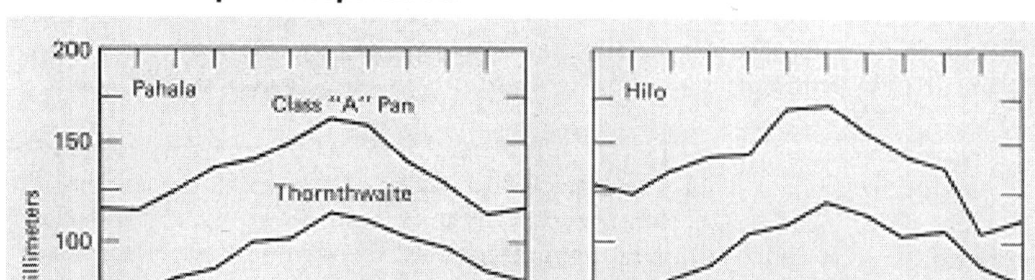

Figure: *Monthly estimated potential evapotranspiration and measured pan evaporation for two locations in Hawaii, Hilo and Pahala.*

Potential evapotranspiration (PET) is the amount of water that would be evaporated and transpired if there were sufficient water available. This demand incorporates the energy available for evaporation and the ability of the lower atmosphere to transport evaporated moisture away from the land surface. PET is higher in the summer, on less cloudy days, and closer to the equator, because of the higher levels of solar radiation that provides the energy for evaporation. PET is also higher on windy days because the evaporated moisture can be quickly moved from the ground or plant surface, allowing more evaporation to fill its place.

PET is expressed in terms of a depth of water, and can be graphed during the year.

Potential evapotranspiration is usually measured indirectly, from other climatic factors, but also depends on the surface type, such as free water (for lakes and oceans), the soil type for bare soil, and the vegetation. Often a value for the potential evapotranspiration is calculated at a nearby climate station on a reference surface, conventionally short grass. This value is called the reference evapotranspiration, and can be converted to a potential evapotranspiration by multiplying with a surface coefficient. In agriculture, this is called a crop coefficient. The difference between potential evapotranspiration and precipitation is used in irrigation scheduling.

Average annual PET is often compared to average annual precipitation, P. The ratio of the two, P/PET, is the aridity index.

Transpiration - Factors Affecting Rates of Transpiration

PLANT PARAMETERS – These plant parameters help plants control rates of transpiration by serving as forms of resistance to water movement out of the plant.

Did you know that an acre of corn can transpire up to 400,000 gallons of water in one growing season?

Stomata are the only way plants can control transpiration rates in the short-term.

Sun leaves have much thicker cuticles than shade leaves causing slower rates of transpiration.

Stomata – Stomata are pores in the leaf that allow gas exchange where water vapor leaves the plant and carbon dioxide enters. Special cells called guard cells control each pore's opening or closing. When stomata are open, transpiration rates increase; when they are closed, transpiration rates decrease.

Boundary Layer – The boundary layer is a thin layer of still air hugging the surface of the leaf. This layer of air is not moving. For transpiration to occur, water vapor leaving the stomata must diffuse through this motionless layer to reach the atmosphere where the water vapor will be removed by moving air. The larger the boundary layer, the slower the rates of transpiration.

Plants can alter the size of their boundary layers around leaves through a variety of structural features. Leaves that possess many hairs or pubescence will have larger boundary layers; the hairs serve as mini-wind breaks by increasing the layer of still air around the leaf surface and slowing transpiration rates. Some plants possess stomata that are sunken into the leaf surface, dramatically increasing the boundary layer and slowing transpiration. Boundary layers increase as leaf size increases, reducing rates of transpiration as well. For example, plants from desert climates often have small leaves so that their small boundary layers will help cool the leaf with higher rates of transpiration.

Cuticle – The cuticle is the waxy layer present on all above-ground tissue of a plant and serves as a barrier to water movement out of a leaf. Because the cuticle is made of wax, it is very hydrophobic or 'water-repelling'; therefore, water does not move through it very easily. The thicker the cuticle layer on a leaf surface, the slower the transpiration rate. Cuticle thickness varies widely among plant species. In general, plants from hot, dry climates have thicker cuticles than plants from cool, moist climates. In addition, leaves that develop under direct sunlight will have much thicker cuticles than leaves that develop under shade conditions.

Environmental Conditions – Some environmental conditions create the driving force for movement of water out of the plant. Others alter the plant's ability to control water loss.

Windier conditions increase transpiration because the leaf's boundary layer is smaller.

Light levels as low as one thousandth of the sun can cause stomata to open.

Relative Humidity – Relative humidity (RH) is the amount of water vapor in the air compared to the amount of water vapor that air could hold at a given temperature. A hydrated leaf would have a RH near 100%, just as the atmosphere on a rainy day would have. Any reduction in water in the atmosphere creates a gradient for water to move from the leaf to the atmosphere. The lower the RH, the less moist the atmosphere and thus, the greater the driving force for transpiration. When RH is high, the atmosphere contains more moisture, reducing the driving force for transpiration.

Temperature – Temperature greatly influences the magnitude of the driving force for water movement out of a plant rather than having a direct effect on stomata. As temperature increases, the water holding capacity of that air increases sharply. The amount of water does not change, just the ability of that air to hold water. Because warmer air can hold more water, its relative humidity is less than the same air sample at a lower temperature, or it is 'drier air'. Because cooler air holds less water, its relative humidity increases or it is 'moister air'. Therefore, warmer air will increase the driving force for transpiration and cooler air will decrease the driving force for transpiration.

Soil Water – The source of water for transpiration out of the plant comes from the soil. Plants with adequate soil moisture will normally transpire at high rates because the soil provides the water to move through the plant. Plants cannot continue to transpire without wilting if the soil is very dry because the water in the xylem that moves out through the leaves is not being replaced by the soil water. This condition causes the leaf to lose turgor or firmness, and the stomata to close. If this loss of turgor continues throughout the plant, the plant will wilt.

Light – Stomata are triggered to open in the light so that carbon dioxide is available for the light-dependent process of photosynthesis. Stomata are closed in the dark in most plants. Very low levels of light at dawn can cause stomata to open so they can access carbon dioxide for photosynthesis as soon as the sun hits their leaves. Stomata are most sensitive to blue light, the light predominating at sunrise.

Wind – Wind can alter rates of transpiration by removing the boundary layer, that still layer of water vapor hugging the surface of leaves. Wind increases the movement of water from the leaf surface when it reduces the boundary layer, because the path for water to reach the atmosphere is shorter.

Transpiration - What Controls Rates of Transpiration?

How fast does water move through plants? Transpiration rates depend on two major factors: 1) the driving force for water movement from the soil to the atmosphere and 2) the resistances to water movement in the plant.

Driving force: The driving force for transpiration is the difference in water potential between the soil and the atmosphere surrounding the plant. This difference creates a gradient, forcing water to move toward areas with less water. The drier the air around the plant, the greater the driving force is for water to move through the plant and the faster the transpiration rate. The following section, FACTORS AFFECTING RATES OF TRANSPIRATION, expands on how changes in the environment alter this driving force and thus transpiration.

Resistances: There are three major resistances to the movement of water out of a leaf: cuticle resistance, stomata resistance and boundary layer resistance. These resistances slow water movement. The greater any individual resistance is to water movement, the slower the transpiration rate. The following section, FACTORS AFFECTING RATES OF TRANSPIRATION, expands on how changes in the plant alter these resistances and thus transpiration.

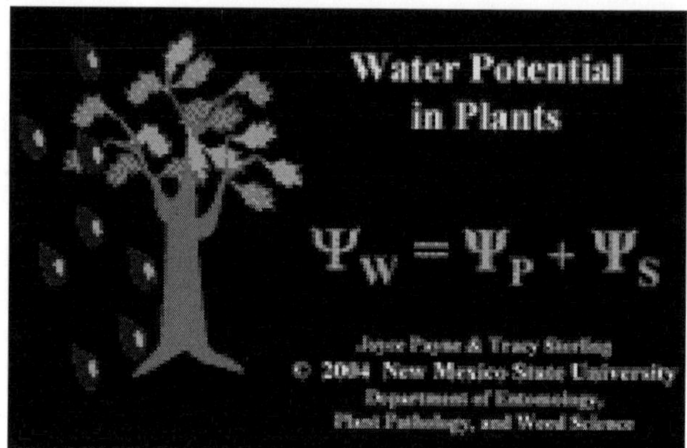

Figure: *NOTE: The PowerPoints may take several seconds to load.

A simple equation describing how these factors alter transpiration is:

Transpiration= [Water potential(leaf)] – [Water potential (atmosphere)]

Resistance

The units for this equation are mols of water lost per leaf area per time (mol/cm2/s). This equation makes predicting rates of transpiration easy. For example, any time the numerator (the value for the driving force) is increased, the rate of transpiration becomes faster and vice versa. Similarly, if the denominator (the value for resistance) increases, this means there is greater resistance and thus,

Transpiration - Major Plant Highlights

Root Detail– The major path for water movement into plants is from soil to roots. Water enters near the tip of a growing root, the same region where root hairs grow. The surface of the root hairs needs to be in close contact with the soil to access soil water. Water diffuses into the root, where it can take at least three different pathways to eventually reach the xylem, the conduit located at the interior of the root that carries the soil water to the leaves. View the next level of this animation to see the possible pathways that water can take across a root.

What path does water take to reach the leaf from the root hair? Once water has entered a root hair, it must move across the cortex and endodermis before it reaches the xylem. Water will take the path of least resistance through a root to reach the xylem.

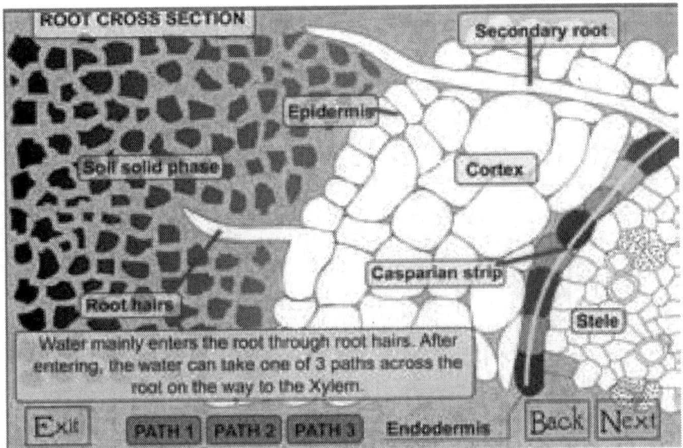

Figure: *Cross section of a plant root*

Water can move across the root via three different pathways. One path is the apoplastic path where the water molecule stays between cells in the cell wall region, never crossing membranes or entering a cell. The other two routes, called cellular pathways, require the water molecule to actually move across a membrane. The first cellular pathway is the transmembrane path where water moves from cell to cell across membranes; it will leave one cell by traversing its membrane and will re-enter another cell by crossing its membrane. The second cellular path is the symplastic path which takes the water molecule from cell to cell using the intercellular connections called the plasmodesmata which are membrane connections between adjacent cells. Regardless of the pathway, once the water molecule has traversed the cortex, it must now cross the endodermis. The endodermis is a layer of cells with a waxy inlay or mortar called the Casparian strip that stops water movement between cells. At this point, water is forced to move through the membranes of endodermal cells, creating a sieving effect. Once in the endodermal cells, the water freely enters the xylem cells where it joins the fast moving column of water or transpiration stream, headed to the leaves.

Xylem Details– The xylem is probably the longest part of the pathway that water takes on its way to the leaves of a plant. It is also the path of least resistance, with about a billion times less resistance than cell to cell transport of water. Xylem cells are called tracheids (cells with narrower diameters) or vessels (cells with wider diameters). Their cell walls contain cellulose and lignin making them extremely rigid. Xylem cells contain no membranes and are considered dead. These cells overlap to create a series of pathways that water can take as it heads to the leaves. There is no single column of xylem cells carrying water.

Plants are most susceptible to cavitation when transpiration rates are extremely high.

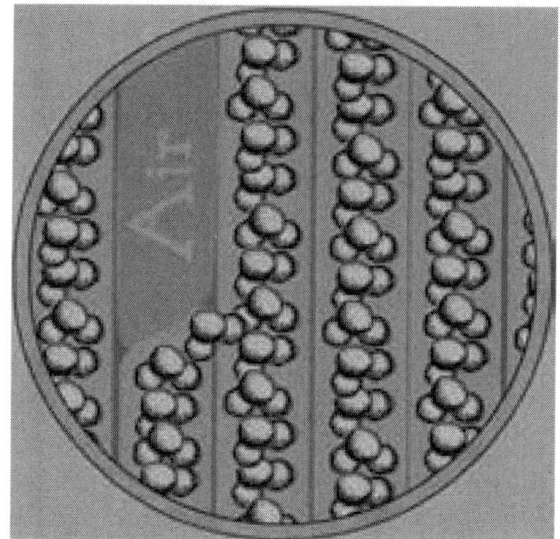

Figure: *Picture of cavitation*

Webster's Dictionary says a stoma (singular form of stomata) is a small, simple opening.

How do stomata open? Stomata sense environmental cues, like light, to open. These cues start a series of reactions that cause their guard cells to fill with water. Let's follow a scenario where the sun is rising and a cotton plant is signaled to open its stomata:

1. Signal received: The blue light at dawn is the signal that is recognised by a receptor on the guard cell.
2. The receptor signals the H^+-ATPases on the guard cell's plasma membrane to start pumping protons (H^+) out of the guard cell. This loss of positive charge creates a negative charge in the cell.
3. Potassium ions (K^+) enter the guard cell through channels in the membrane, moving toward its more negative interior.

4. As the potassium ions accumulate in the guard cell, the osmotic pressure is lowered.
5. A lower osmotic pressure attracts water to enter the cell.
6. As water enters the guard cell, its hydrostatic pressure increases.

The pressure causes the shape of the guard cells to change and a pore is formed, allowing gas exchange.

Cavitation– Cavitation is the filling of a xylem vessel or tracheid with air. It is also known as an 'embolism' or 'air-lock'. Remember that during transpiration, the column of water is being pulled out of the plant by evaporation at the leaf cell surface. When this 'pulling' of water out of the plant becomes greater than the ability of the water molecules to stay together, the column of water will break. Using sound-sensing equipment, one can actually hear a 'click' when the water molecules split from one another. Unique structural characteristics help the plant contain the air bubble so that it does not totally disrupt water movement up the plant.

Plants are particularly sensitive to cavitation during the hottest part of the day when there is not enough water available from the soil to keep up with the demand for water while it is evaporating off the leaf surface. Cavitation also occurs under freezing conditions. Because the solubility of gas in ice is very low, gas comes out of solution when the xylem sap freezes. Freezing of xylem sap is a problem in the spring when the ice thaws, leaving a bubble in a xylem vessel. These bubbles can block water transport and cause water deficit in leaves.

Plants avoid cavitation or minimize its damage through several mechanisms:

1. Xylem cells possess pits or tiny holes that allow liquid water transport, but do not allow the gas bubble to escape; this structural characteristic helps keep the gas bubble in one cell, so the other xylem cells can continue to transport water up the plant.
2. Water will detour around any xylem cell containing an air bubble through the pits as well.
3. The gas bubble will re-dissolve into liquid water when the pulling of water through the xylem is reduced, such as during the night when water is not being pulled out of the leaf via transpiration because the stomata are closed.

4. Xylem cells with narrower diameters (tracheids) compared to those with wider diameters (vessels) avoid cavitation because the column of water in a cell with a narrow diameter is better able to resist bubble formation or rupture.

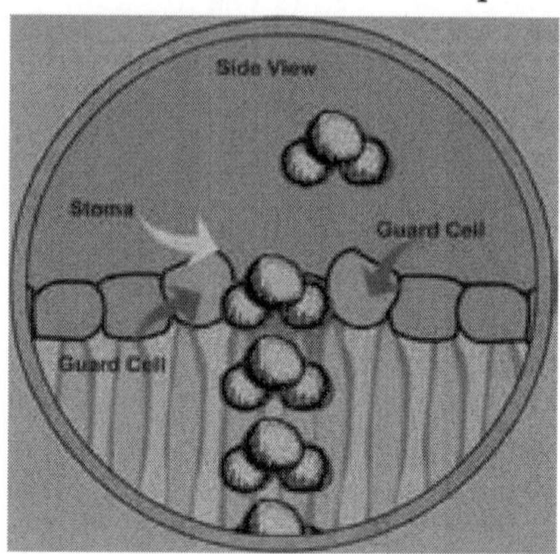

Figure: *Picture of water molecules leaving stomata - side view*

Stomata Details– The stomata are the primary control mechanisms that plants use to reduce water loss and they are able to do so quickly. Stomata are sensitive to the environmental cues that trigger the stomata to open or close. The major role of stomata is to allow carbon dioxide entry to drive photosynthesis and at the same time allow the exit of water as it evaporates, cooling the leaf. Two specialised cells called 'guard cells' make up each stoma (stoma is singular for stomata). Plants have many stomata (up to 400 per mm^2) on their leaf surfaces and they are usually on the lower surface to minimize water loss.

SIDE VIEW OF STOMATA– Environmental cues that affect stomata opening and closing are light, water, temperature, and the concentration of CO2 within the leaf. Stomata will open in the light and close in the dark. However, stomata can close in the middle of the day if water is limiting, CO2 accumulates in the leaf, or the temperature is too hot. If the plant lacks water, stomata will close because there will not be enough water to create pressure in the guard cells for stomatal opening; this response helps the plant conserve water.

If the leaf's internal concentration of CO2 increases, the stomata are signaled to close because respiration is releasing more CO2 than photosynthesis is using. There is no need to keep the stomata open and lose water if photosynthesis is not functioning. Alternatively, if the leaf's CO2 concentration is low, the stomata will stay open to continue fuelling photosynthesis. High temperatures will also signal stomata to close.

How do stomata open? Stomata sense environmental cues, like light, to open. These cues start a series of reactions that cause their guard cells to fill with water. Let's follow a scenario where the sun is rising and a cotton plant is signaled to open its stomata:

1. Signal received: The blue light at dawn is the signal that is recognised by a receptor on the guard cell.
2. The receptor signals the H^+-ATPases on the guard cell's plasma membrane to start pumping protons (H^+) out of the guard cell. This loss of positive charge creates a negative charge in the cell.
3. Potassium ions (K^+) enter the guard cell through channels in the membrane, moving toward its more negative interior.
4. As the potassium ions accumulate in the guard cell, the osmotic pressure is lowered.
5. A lower osmotic pressure attracts water to enter the cell.
6. As water enters the guard cell, its hydrostatic pressure increases.
7. The pressure causes the shape of the guard cells to change and a pore is formed, allowing gas exchange.

Cavitation– Cavitation is the filling of a xylem vessel or tracheid with air. It is also known as an 'embolism' or 'air-lock'. Remember that during transpiration, the column of water is being pulled out of the plant by evaporation at the leaf cell surface. When this 'pulling' of water out of the plant becomes greater than the ability of the water molecules to stay together, the column of water will break. Using sound-sensing equipment, one can actually hear a 'click' when the water molecules split from one another. Unique structural characteristics help the plant contain the air bubble so that it does not totally disrupt water movement up the plant.

Plants are particularly sensitive to cavitation during the hottest part of the day when there is not enough water available from the soil to keep up with the demand for water while it is evaporating off the leaf surface. Cavitation also occurs under freezing conditions. Because the solubility of gas in ice is very low, gas comes out of solution when the xylem sap freezes. Freezing of xylem sap is a problem in the spring when the ice thaws, leaving a bubble in a xylem vessel. These bubbles can block water transport and cause water deficit in leaves.

Plants avoid cavitation or minimize its damage through several mechanisms:

1. Xylem cells possess pits or tiny holes that allow liquid water transport, but do not allow the gas bubble to escape; this structural characteristic helps keep the gas bubble in one cell, so the other xylem cells can continue to transport water up the plant.
2. Water will detour around any xylem cell containing an air bubble through the pits as well.
3. The gas bubble will re-dissolve into liquid water when the pulling of water through the xylem is reduced, such as during the night when water is not being pulled out of the leaf via transpiration because the stomata are closed.
4. Xylem cells with narrower diameters (tracheids) compared to those with wider diameters (vessels) avoid cavitation because the column of water in a cell with a narrow diameter is better able to resist bubble formation or rupture.

Stomata Details– The stomata are the primary control mechanisms that plants use to reduce water loss and they are able to do so quickly. Stomata are sensitive to the environmental cues that trigger the stomata to open or close. The major role of stomata is to allow carbon dioxide entry to drive photosynthesis and at the same time allow the exit of water as it evaporates, cooling the leaf. Two specialised cells called 'guard cells' make up each stoma (stoma is singular for stomata). Plants have many stomata (up to 400 per mm^2) on their leaf surfaces and they are usually on the lower surface to minimize water loss.

Side View of Stomata– Environmental cues that affect stomata opening and closing are light, water, temperature, and the concentration of CO2 within the leaf. Stomata will open in the light and close in the dark. However, stomata can close in the middle of the day if water is limiting, CO2 accumulates in the leaf, or the temperature is too hot. If the plant lacks water, stomata will close because there will not be enough water to create pressure in the guard cells for stomatal opening; this response helps the plant conserve water.

If the leaf's internal concentration of CO2 increases, the stomata are signaled to close because respiration is releasing more CO2 than photosynthesis is using. There is no need to keep the stomata open and lose water if photosynthesis is not functioning. Alternatively, if the leaf's CO2 concentration is low, the stomata will stay open to continue fuelling photosynthesis. High temperatures will also signal stomata to close.

High temperatures will increase the water loss from the leaf. With less water available, guard cells can become flaccid and close. Another effect of high temperatures is that respiration rates rise above photosynthesis rates causing an increase of CO2 in the leaves; high internal CO2 will cause stomata to close as well. Remember that some plants may open their stomata under high temperatures so that transpiration will cool the leaves.

Top View of Stomata –

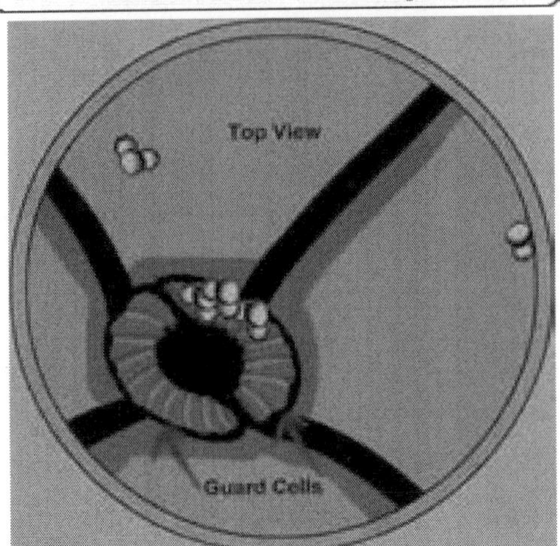

Figure: *Picture of open stomata - top view*

Open Stomata– When stomata are signaled to open, potassium ions (K+) enter the guard cells. This causes water to enter down its water potential gradient, creating a hydrostatic pressure in the guard cell that changes the shape of the stoma. Guard cells expand on the outer edges of the stoma, but not on the inner side, resulting in kidney-shaped cells and an opening or pore between the two guard cells for gas exchange. Kidney-shaped guard cells are characteristic of dicots; however, many plant (e.g. grasses, other monocots) have dumbbell-shaped stomata. The shape taken by the guard cells is dependent on cellulose microfibrils that fan out radially from the pore, somewhat similar to radial tires. The cellulose microfibrils are rigid and do not stretch when water has entered the cell. The cell walls surrounding the stomatal opening are thickened, preventing that side of the guard cell from expanding. Therefore, when pressure in the cell increases due to water entry, guard cell does not widen, but rather the outer edge stretches disproportionately more than the inner edge. This unequal stretching allows the pore to form between the two guard cells.

> Many plant leaves have stomata only on their lower surface to help avoid water loss.

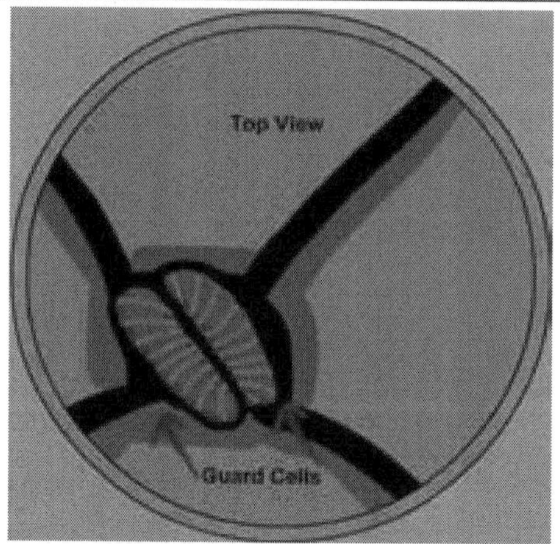

Figure: *Picture of closed stomata - top view*

Closed Stomata– Stomata must be open for the plant to photosynthesize; however, open stomata present a risk of losing too much water through transpiration. Stomata close when the guard cells lose water and become flaccid. This occurs because potassium ions move back out of the guard cell, followed by water that lowers the pressure in the.

5

Raising Water

Root pressure is osmotic pressure within the cells of a root system that causes sap to rise through a plant stem to the leaves.

Root pressure occurs in the xylem of some vascular plants when the soil moisture level is high either at night or when transpiration is low during the day. When transpiration is high, xylem sap is usually under tension, rather than under pressure, due to transpirational pull. At night in some plants, root pressure causes guttation or exudation of drops of xylem sap from the tips or edges of leaves. Root pressure is studied by removing the shoot of a plant near the soil level. Xylem sap will exude from the cut stem for hours or days due to root pressure. If a pressure gauge is attached to the cut stem, the root pressure can be measured.

Root pressure is caused by active distribution of mineral nutrient ions into the root xylem. Without transpiration to carry the ions up the stem, they accumulate in the root xylem and lower the water potential. Water then diffuses from the soil into the root xylem due to osmosis. Root pressure is caused by this accumulation of water in the xylem pushing on the rigid cells. Root pressure provides a force, which pushes water up the stem, but it is not enough to account for the movement of water to leaves at the top of the tallest trees. The maximum root pressure measured in some plants can raise water only to about 7 metres, and the tallest trees are over 100 metres tall.

Role of Endodermis

The endodermis in the root is important in the development of root pressure. The endodermis is a single layer of cells between the cortex and the pericycle. These cells allow water movement until it reaches the Casparian strip, made of suberin, a waterproof substance. The Casparian strip prevents mineral nutrient ions from moving passively through the endodermal cell walls. Water and ions move in these cell walls via the apoplast pathway. Ions outside the endodermis must be actively transported across an endodermal cell membrane to enter or exit the endodermis. Once inside the endodermis, the ions are in the symplast pathway.

They cannot diffuse back out again but can move from cell to cell via plasmodesmata or be actively transported into the xylem. Once in the xylem vessels or tracheids, ions are again in the apoplast pathway. Xylem vessels and tracheids transport water up the plant but lack cell membranes. The Casparian strip substitutes for their lack of cell membranes and prevents accumulated ions from diffusing passively in apoplast pathway out of the endodermis. The ions accumulating interior to the endodermis in the xylem create a water potential gradient and by osmosis, water diffuses from the moist soil, across the cortex, through the endodermis and into the xylem.

Importance

Root pressure can transport water and dissolved mineral nutrients from roots through the xylem to the tops of relatively short plants when transpiration is low or zero. The maximum root pressure measured is about 0.6 megapascals but some species never generate any root pressure. The main contributor to the movement of water and mineral nutrients upward in vascular plants is considered to be the transpirational pull.

However, sunflower plants grown in 100% relative humidity grew normally and accumulated the same amount of mineral nutrients as plants in normal humidity, which had a transpiration rate 10 to 15 times the plants in 100% humidity. Thus, transpiration may not be as important in upward mineral nutrient transport in relatively short plants as often assumed. Xylem vessels sometimes empty over winter. Root pressure may be important in refilling the xylem vessels. However, in some species vessels refill without root pressure.

Root pressure is often high in some deciduous trees before they leaf out. Transpiration is minimal without leaves, and organic solutes are being mobilized so decrease the xylem water potential. Sugar maple accumulates high concentrations of sugars in its xylem early in the spring, which is the source of maple sugar. Some trees “bleed” xylem sap profusely when their stems are pruned in late winter or early spring, e.g. maple and elm. Such bleeding is similar to root pressure only sugars, rather than ions, may lower the xylem water potential. In the unique case of maple trees, sap bleeding is caused by changes in stem pressure and not root pressure.

Staining Science: Capillary Action of Dyed Water in Plants

Have you ever heard someone say, "That plant is thirsty," or "Give that plant a drink of water."? We know that all plants need water to survive, even bouquets of cut flowers and plants living in deserts. But have you ever thought about how water moves within the plant? In this activity, you'll put carnations in dyed water to figure out where the water goes. Where do you think the dyed water will travel, and what will this tell you about how the water moves in the cut flowers?

Background

Plants use water to keep their roots, stems, leaves and flowers healthy as well as prevent them from drying and wilting. The water is also used to carry dissolved nutrients throughout the plant.

Most of the time, plants get their water from the ground. This means it has to transport the water from its roots up and throughout the rest of the plant. How does it do this? Water moves through the plant by means of capillary action. Capillary action occurs when the forces binding a liquid together (cohesion and surface tension) and the forces attracting that bound liquid to another surface (adhesion) are greater than the force of gravity. Through these binding and surface forces, the plant's stem basically sucks up water—almost like drinking through a straw!

A simple way of observing capillary action is to take a teaspoon of water and gently pour it in a pool on a countertop. You'll notice that the water stays together in the pool, rather than flattening out across the countertop. (This happens because of cohesion and surface tension.) Now gently dip the corner of a paper towel in the pool of water. The water adheres to the paper and "climbs" up the paper towel. This is called capillary action.

Materials

- Water
- Measuring cup
- Glass cup or vase
- Blue or red food colour

- Several white carnations (at least three). Tip: Fresher flowers work better than older ones
- Knife
- Camera (optional)

Preparation

- Measure a half cup of water and pour it into the glass or vase.
- Add 20 drops of food colour to the water in the glass.
- With the help of an adult, use a knife to cut the bottom stem tips of several (at least three) white carnations at a 45-degree angle. Tip: Be sure not to use scissors, they will crush the stems, reducing their ability to absorb water. Also, shorter stems work better than longer ones.
- Place the carnations in the dyed water. As you do this, use the stems of the carnations to stir the water until the dye has fully dissolved.

Procedure

- Observe the flowers immediately after you put them in the water. If you have a camera, take a picture of the flowers.
- Observe the flowers two, four, 24, 48 and 72 hours after you put them in the dyed water. Be sure to also observe their stems, especially the bumps where the leaves branch from the stem and it is lighter gre en (it may be easier to see the dye here). If you have a camera, take pictures of the flowers and stems at these time points.
- *How did the flowers look after two hours? What about after four, 24, 48 and 72 hours? How did their appearance change over this time period?*
- *What does the flowers' change in appearance tell you about how water moves through them?*
- Extra: *In this activity, you used carnations, but do you think you'd see the same results with other flowers and plants?* Try this activity with another white flower— a daisy, for instance— or a plant that is mostly stem, such as a stalk of celery.
- Extra: Try doing this activity again but use higher or lower concentrations of food colour, such as one half, twice, four times or 10 times as much; be sure to mix each dye amount with the same amount of water. *What happens if you increase or decrease the concentration of food colour in the water?*

- Extra: *How would you make a multicolour carnation?* Tip: You could try (1) leaving the flower for a day in one colour of water and then putting it in another colour of water for a second day or (2) splitting the end of the stem in two and immersing each half in a different colour of water.

Observations and Results

When you put the flowers in the dyed water, did you see some of the flowers start to show spots of dye after two hours? Did you also see some dye in the stems? After 24 hours did the flowers overall have a coloured hue to them? Did this hue become more pronounced, or darker, after 48 and 72 hours?

Water moves through the plant by means of capillary action. Specifically, the water is pulled through the stem and then makes its way up to the flower. After two hours of being in the dyed water, some flowers should have clearly showed dyed spots near the edges of their petals. The water that has been pulled up undergoes a process called transpiration, which is when the water from leaves and flower petals evaporates. However, the dye it brought along doesn't evaporate, and stays around to colour the flower. The loss of water generates low water pressure in the leaves and petals, causing more coloured water to be pulled through the stem. By 24 hours the flowers should have gained an overall dyed hue, which darkened a little over time. The stems should have also become slightly dyed in places, particularly where the leaves branch off.

How does Water Move through Plants to Get to the Top of Tall Trees?

Why do Plants Need So Much Water?

Water is the most limiting abiotic (non-living) factor to plant growth and productivity, and a principal determinant of vegetation distributions worldwide. Since antiquity, humans have recognised plants' thirst for water as evidenced by the existence of irrigation systems at the beginning of recorded history. Water's importance to plants stems from its central role in growth and photosynthesis, and the distribution of organic and inorganic molecules. Despite this dependence, plants retain less than 5% of the water absorbed by roots for cell expansion and plant growth. The remainder passes through plants directly into the atmosphere, a process referred to as transpiration. The amount of water lost via transpiration can be incredibly high; a single irrigated corn plant growing in Kansas can use 200 L of water during a typical summer, while some large rainforest trees can use nearly 1200 L of water in a single day!

If water is so important to plant growth and survival, then why would plants waste so much of it? The answer to this question lies in another process vital to plants — photosynthesis. To make sugars, plants must absorb carbon dioxide (CO_2) from the atmosphere through small pores in their leaves called stomata. However, when stomata open, water is lost to the atmosphere at a prolific rate relative to the small amount of CO_2 absorbed; across plant species an average of 400 water molecules are lost for each CO_2 molecule gained. The balance between transpiration and photosynthesis forms an essential compromise in the existence of plants; stomata must remain open to build sugars but risk dehydration in the process.

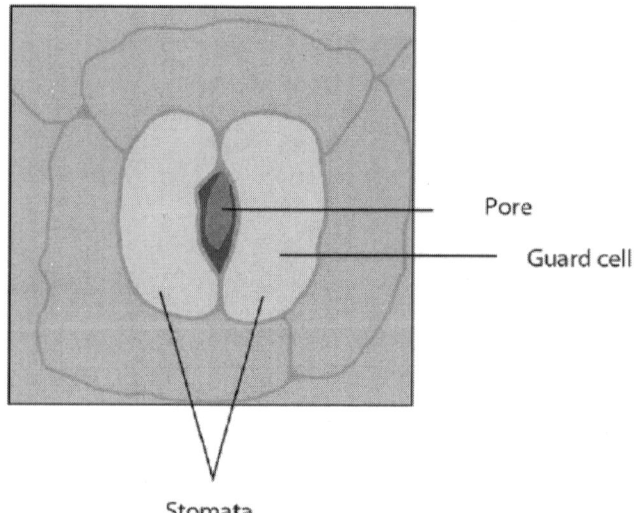

Figure: *Rendering of an open stoma on the surface of a tobacco leaf.*

Stomata are pores found on the leaf surface that regulate the exchange of gases between the leaf's interior and the atmosphere. Stomatal closure is a natural response to darkness or drought as a means of conserving water.

From the Soil into the Plant

Essentially all of the water used by land plants is absorbed from the soil by roots. A root system consists of a complex network of individual roots that vary in age along their length. Roots grow from their tips and initially produce thin and non-woody fine roots. Fine roots are the most permeable portion of a root system, and are thought to have the greatest ability to absorb water, particularly in herbaceous (i.e., non-woody) plants (McCully 1999). Fine roots can be covered by root hairs that significantly increase the absorptive surface area and improve contact between roots and the soil. Some plants also improve water uptake by establishing symbiotic relationships with mycorrhizal fungi, which functionally increase the total absorptive surface area of the root system.

Figure: *Root hairs often form on fine roots and improve water absorption by increasing root surface area and by improving contact with the soil.*

Roots of woody plants form bark as they age, much like the trunks of large trees. While bark formation decreases the permeability of older roots they can still absorb considerable amounts of water (MacFall *et al.* 1990, Chung & Kramer 1975). This is important for trees and shrubs since woody roots can constitute ~99% of the root surface in some forests (Kramer & Bullock 1966).

Roots have the amazing ability to grow away from dry sites toward wetter patches in the soil — a phenomenon called hydrotropism. Positive hydrotropism occurs when cell elongation is inhibited on the humid side of a root, while elongation on the dry side is unaffected or slightly stimulated resulting in a curvature of the root and growth toward a moist patch (Takahashi 1994).

The root cap is most likely the site of hydrosensing; while the exact mechanism of hydrotropism is not known, recent work with the plant model *Arabidopsis* has shed some light on the mechanism at the molecular level.

Roots of many woody species have the ability to grow extensively to explore large volumes of soil. Deep roots (>5 m) are found in most environments (Canadell *et al.* 1996, Schenk & Jackson 2002) allowing plants to access water from permanent water sources at substantial depth. Roots from the Shepard's tree (*Boscia albitrunca*) have been found growing at depths 68 m in the central Kalahari, while those of other woody species can spread laterally up to 50 m on one side of the plant (Schenk & Jackson 2002). Surprisingly, most arid-land plants have very shallow root systems, and the deepest roots consistently occur in climates with strong seasonal precipitation (i.e., Mediterranean and monsoonal climates).

Through the Plant into the Atmosphere

Water flows more efficiently through some parts of the plant than others. For example, water absorbed by roots must cross several cell layers before entering the specialised water transport tissue (referred to as xylem). These cell layers act as a filtration system in the root and have a much greater resistance to water flow than the xylem, where transport occurs in open tubes. Imagine the difference between pushing water through numerous coffee filters versus a garden hose. The relative ease with which water moves through a part of the plant is expressed quantitatively using the following equation:

$$\text{Flow} = \Delta\Psi / R,$$

which is analogous to electron flow in an electrical circuit described by Ohm's law equation:

$$i = V / R,$$

where R is the resistance, i is the current or flow of electrons, and V is the voltage. In the plant system, V is equivalent to the water potential difference driving flow ($\Delta\Psi$) and i is equivalent to the flow of water through/across a plant segment. Using these plant equivalents, the Ohm's law analogy can be used to quantify the hydraulic conductance (i.e., the inverse of hydraulic R) of individual segments (i.e., roots, stems, leaves) or the whole plant (from soil to atmosphere).

Upon absorption by the root, water first crosses the epidermis and then makes its way toward the centre of the root crossing the cortex and endodermis before arriving at the xylem. Along the way, water travels in cell walls (apoplastic pathway) and/or through the inside of cells (cell to cell pathway, C-C) (Steudle 2001).

At the endodermis, the apoplastic pathway is blocked by a gasket-like band of suberin — a waterproof substance that seals off the route of water in the apoplast forcing water to cross via the C-C pathway.

Because water must cross cell membranes (e.g., in the cortex and at apoplastic barriers), transport efficiency of the C-C pathway is affected by the activity, density, and location of water-specific protein channels embedded in cell membranes (i.e., aquaporins).

Much work over the last two decades has demonstrated how aquaporins alter root hydraulic resistance and respond to abiotic stress, but their exact role in bulk water transport is yet unresolved.

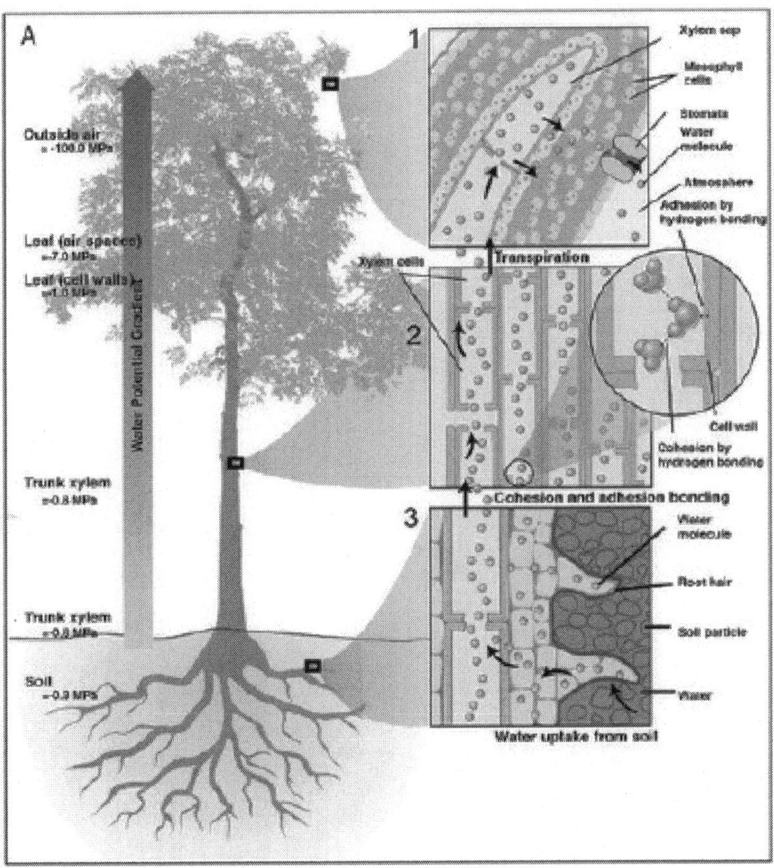

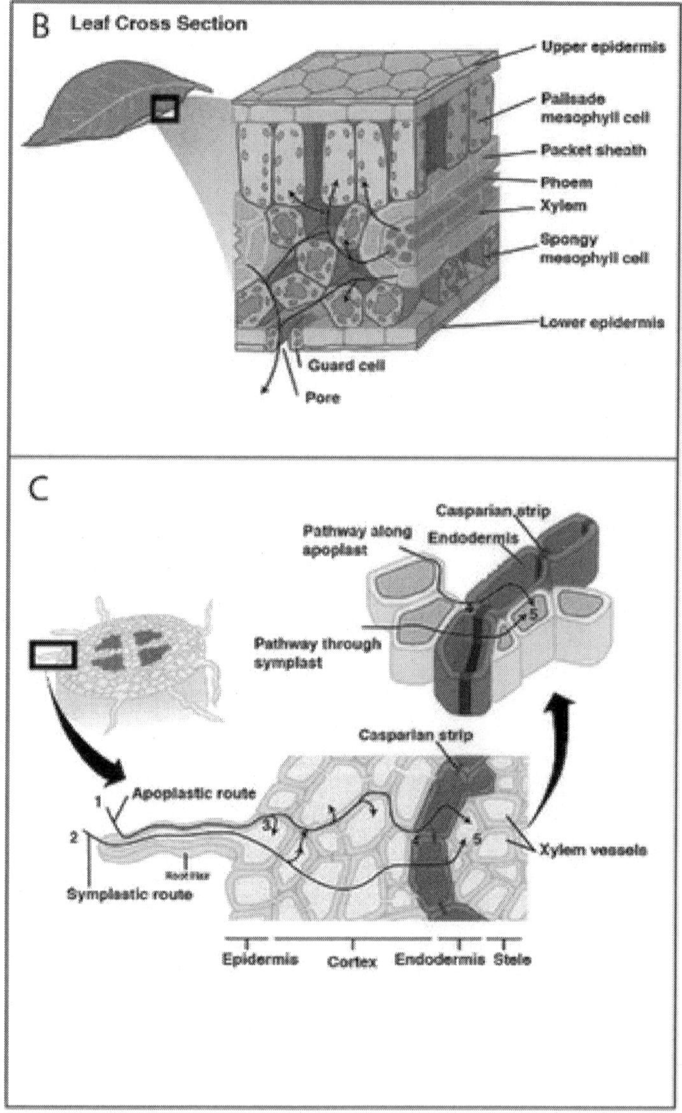

Figure: *Representation of the water transport pathways along the soil-plant-atmosphere continuum (SPAC).*

(A) Water moves from areas of high water potential (i.e. close to zero in the soil) to low water potential (i.e., air outside the leaves). Details of the Cohesion-Tension mechanism are illustrated with the inset panels (A), where tension is generated by the evaporation of water molecules during leaf transpiration (1) and is transmitted down the continuous, cohesive water columns (2) through the xylem and out the roots to the soil (3). The pathways for water movement out of the leaf veins and through the stomata (B) and across the fine roots (C) are detailed and illustrate both symplastic and apoplastic pathways.

Once in the xylem tissue, water moves easily over long distances in these open tubes. There are two kinds of conducting elements (i.e., transport tubes) found in the xylem: 1) tracheids and 2) vessels. Tracheids are smaller than vessels in both diameter and length, and taper at each end. Vessels consist of individual cells, or "vessel elements", stacked end-to-end to form continuous open tubes, which are also called xylem conduits. Vessels have diameters approximately that of a human hair and lengths typically measuring about 5 cm although some plant species contain vessels as long as 10 m. Xylem conduits begin as a series of living cells but as they mature the cells commit suicide (referred to as programmed cell death), undergoing an ordered deconstruction where they lose their cellular contents and form hollow tubes. Along with the water conducting tubes, xylem tissue contains fibers which provide structural support, and living metabolically-active parenchyma cells that are important for storage of carbohydrates, maintenance of flow within a conduit, and radial transport of water and solutes.

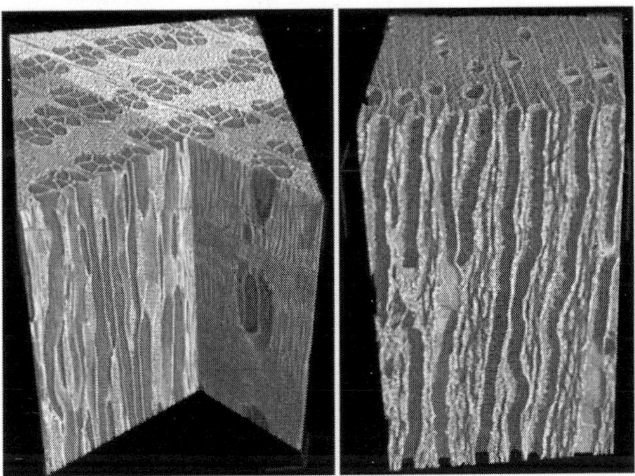

Figure: *Three dimensional reconstructions of xylem imaged at the Ghent microCT facility.*

Differences in xylem structure and conduit distributions can be seen between *Ulmus americana* (left) and *Fraxinus americana* (right) xylem.

When water reaches the end of a conduit or passes laterally to an adjacent one, it must cross through pits in the conduit cell walls. Bordered pits are cavities in the thick secondary cell walls of both vessels and tracheids that are essential components in the water-transport system of higher plants. The pit membrane, consisting of a modified primary cell wall and middle lamella, lies at the centre of each pit, and allows water to pass between xylem conduits while limiting the spread of air bubbles (i.e., embolism) and xylem-dwelling pathogens. Thus, pit membranes function as safety valves in the plant water transport system. Averaged across a wide range of species, pits account for >50% of total xylem hydraulic resistance. The structure of pits varies dramatically across species, with large differences evident in the amount of conduit wall area covered by pits, and in the porosity and thickness of pit membranes.

This features wider conduits from flowering plants (top), a cartoon reconstruction of vessels, tracheids and their pit membranes (middle), which are also shown in SEM images (bottom).

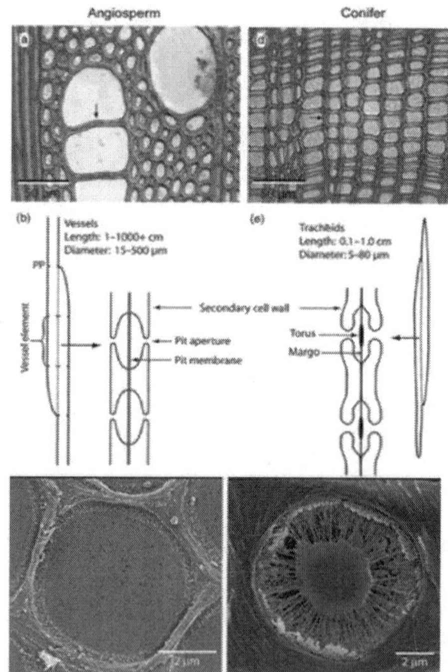

Figure: *Comparison of different types of wood from flowering and cone-bearing plants.*

After travelling from the roots to stems through the xylem, water enters leaves via petiole (i.e., the leaf stalk) xylem that branches off from that in the stem. Petiole xylem leads into the mid-rib (the main thick vein in leaves), which then branch into progressively smaller veins that contain tracheids and are embedded in the leaf mesophyll. In dicots, minor veins account for the vast majority of total vein length, and the bulk of transpired water is drawn out of minor veins (Sack & Holbrook 2006, Sack & Tyree 2005). Vein arrangement, density, and redundancy are important for distributing water evenly across a leaf, and may buffer the delivery system against damage (i.e., disease lesions, herbivory, air bubble spread). Once water leaves the xylem, it moves across the bundle sheath cells surrounding the veins. It is still unclear the exact path water follows once it passes out of the xylem through the bundle sheath cells and into the mesophyll cells, but is likely dominated by the apoplastic pathway during transpiration (Sack & Holbrook 2005).

Figure: *An example of a venation pattern to illustrate the hydraulic pathway from petiole xylem into the leaf cells and out the stomata.*

Mechanism Driving Water Movement in Plants

Unlike animals, plants lack a metabolically active pump like the heart to move fluid in their vascular system. Instead, water movement is passively driven by pressure and chemical potential gradients. The bulk of water absorbed and transported through plants is moved by negative pressure generated by the evaporation of water from the leaves (i.e., transpiration) — this process is commonly referred to as the Cohesion-Tension (C-T) mechanism. This system is able to function because water is "cohesive" — it sticks to itself through forces generated by hydrogen bonding. These hydrogen bonds allow water columns in the plant to sustain substantial tension (up to 30 MPa when water is contained in the minute capillaries found in plants), and helps explain how water can be transported to tree canopies 100 m above the soil surface.

The tension part of the C-T mechanism is generated by transpiration. Evaporation inside the leaves occurs predominantly from damp cell wall surfaces surrounded by a network of air spaces. Menisci form at this air-water interface, where apoplastic water contained in the cell wall capillaries is exposed to the air of the sub-stomatal cavity. Driven by the sun's energy to break the hydrogen bonds between molecules, water evaporates from menisci, and the surface tension at this interface pulls water molecules to replace those lost to evaporation. This force is transmitted along the continuous water columns down to the roots, where it causes an influx of water from the soil. Scientists call the continuous water transport pathway the Soil Plant Atmosphere Continuum (SPAC).

Stephen Hales was the first to suggest that water flow in plants is governed by the C-T mechanism; in his 1727 book Hales states "for without perspiration the [water] must stagnate, notwithstanding the sap-vessels are so curiously adapted by their exceeding fineness, to raise [water] to great heights, in a reciprocal proportion to their very minute diameters." More recently, an evaporative flow system based on negative pressure has been reproduced in the lab for the first time by a 'synthetic tree' (Wheeler & Stroock 2008).

When solute movement is restricted relative to the movement of water (i.e., across semipermeable cell membranes) water moves according to its chemical potential (i.e., the energy state of water) by osmosis — the diffusion of water. Osmosis plays a central role in the movement of water between cells and various compartments within plants. In the absence of transpiration, osmotic forces dominate the movement of water into roots. This manifests as root pressure and guttation — a process commonly seen in lawn grass, where water droplets form at leaf margins in the morning after conditions of low evaporation. Root pressure results when solutes accumulate to a greater concentration in root xylem than other root tissues. The resultant chemical potential gradient drives water influx across the root and into the xylem. No root pressure exists in rapidly transpiring plants, but it has been suggested that in some species root pressure can play a central role in the refilling of non-functional xylem conduits particularly after winter.

Disruption of Water Movement

Water transport can be disrupted at many points along the SPAC resulting from both biotic and abiotic factors. Root pathogens (both bacteria and fungi) can destroy the absorptive surface area in the soil, and similarly foliar pathogens can eliminate evaporative leaf surfaces, alter stomatal function, or disrupt the integrity of the cuticle. Other organisms (i.e., insects and nematodes) can cause similar disruption of above and below ground plant parts involved in water transport. Biotic factors responsible for ceasing flow in xylem conduits include: pathogenic organisms and their by-products that plug conduits; plant-derived gels and gums produced in response to pathogen invasion; and tyloses, which are outgrowths produced by living plant cells surrounding a vessel to seal it off after wounding or pathogen invasion.

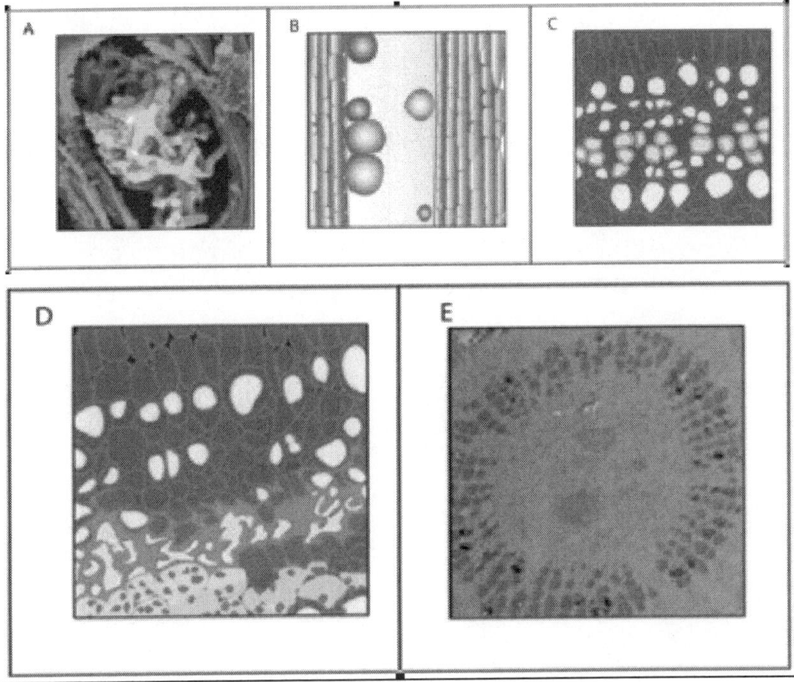

Figure: *Sources of dysfunction in the xylem.*

Left to right: (A) xylem-dwelling pathogens like *Xylella fastidiosa* bacteria; (B) tyloses (plant-derived); (C and D) conduit (in blue) implosion (Brodribb and Holbrook 2005, Pine needle tracheids); and (E) embolized conduits among water filled ones in a frozen plant samples (Choat unpublished figure, Cryo SEM).

Abiotic factors can be equally disruptive to flow at various points along the water transport pathway. During drought, roots shrink and lose contact with water adhering to soil particles — a process that can also be beneficial by limiting water loss by roots to drying soils (i.e., water can flow in reverse and leak out of roots being pulled by drying soil). Under severe plant dehydration, some pine needle conduits can actually collapse as the xylem tensions increase.

Water moving through plants is considered meta-stable because at a certain point the water column breaks when tension becomes excessive — a phenomenon referred to as cavitation. After cavitation occurs, a gas bubble (i.e., embolism) can form and fill the conduit, effectively blocking water movement. Both sub-zero temperatures and drought can cause embolisms. Freezing can induce embolism because air is forced out of solution when liquid water turns to ice. Drought also induces embolism because as plants become drier tension in the water column increases. There is a critical point where the tension exceeds the pressure required to pull air from an empty conduit to a filled conduit across a pit membrane — this aspiration is known as air seeding. An air seed creates a void in the water, and the tension causes the void to expand and break the continuous column. Air seeding thresholds are set by the maximum pore diameter found in the pit membranes of a given conduit.

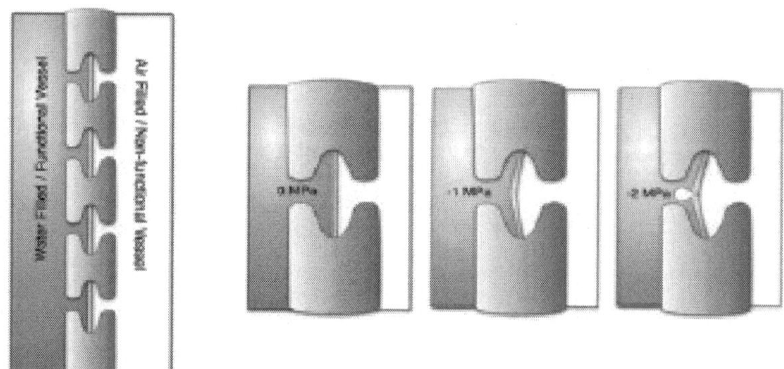

Figure: *Air seeding mechanism.*

Demonstrates how increasing tension in a functional water filled vessel eventually reaches a threshold where an air seed is pulled across a pit membrane from an embolized conduit. Air is seeded into the functional conduit only after the threshold pressure is reached.

Fixing the Problem

Failure to re-establish flow in embolized conduits reduces hydraulic capacity, limits photosynthesis, and results in plant death in extreme cases. Plants can cope with emboli by diverting water around blockages via pits connecting adjacent functional conduits, and by growing new xylem to replace lost hydraulic capacity.

Some plants possess the ability to repair breaks in the water columns, but the details of this process in xylem under tension have remained unclear for decades. Brodersen *et al.* (2010) recently visualized and quantified the refilling process in live grapevines (*Vitis vinifera* L.) using high resolution x-ray computed tomography (a type of CAT scan). Successful vessel refilling was dependent on water influx from living cells surrounding the xylem conduits, where individual water droplets expanded over time, filled vessels, and forced the dissolution of entrapped gas.

The capacity of different plants to repair compromised xylem vessels and the mechanisms controlling these repairs are currently being investigated.

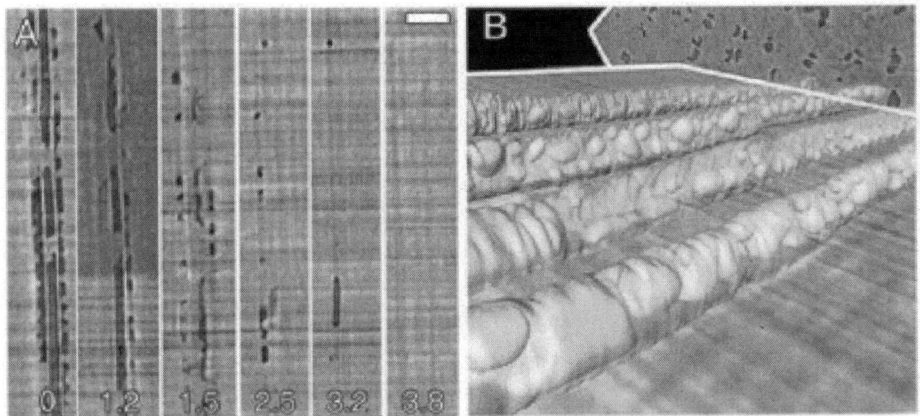

Figure: *Embolism repair documented in grapevines (*Vitis vinifera *L.) with X-ray micro-CT at the ALS facility at Lawrence Berkeley National Lab CA, USA.*

(A) Longitudinal section showing a time series of cavitated vessels refilling in less than 4 hrs; (B) 3D reconstruction of four vessel lumen with water droplets forming on the vessel walls and growing over time to completely fill the embolized conduit.

Transpiration Stream

1. Water is passively transported into the roots and then into the xylem.
2. The forces of cohesion and adhesion cause the water molecules to form a column in the xylem.
3. Water moves from the xylem into the mesophyll cells, evaporates from their surfaces and leaves the plant by diffusion through the stomata.

Figure: *Overview of transpiration.*

In plants, the transpiration stream is the uninterrupted stream of water and solutes which is taken up by the roots and transported via the xylem vessels to the leaves where it evaporates into the air/apoplast-interface of the substomatal cavity. It is driven by capillary action and in some plants by root pressure. The main driving factor is the difference in water potential between the soil and the substomatal cavity caused by transpiration.

Transpiration

Transpiration can be regulated through stomatal closure or opening causing water to move out and evaporate on the leaf, while air flows in. It allows for plants to efficiently transport water up to their highest body organs, regulate the temperature of stem and leaves during heat-generating photosynthesis and it allows for upstream signalling such as the dispersal of an apoplastic alkalinization during local oxidative stress.

A very quick summary of this is:

1. Soil
2. Roots and Root Hair
3. Xylem
4. Leaves
5. Air

Osmosis

The water passes from the soil to the root by osmosis. This is helped by the root hairs' shape – as they are long and thin, they maximize the surface area, meaning that more water can enter into the root hairs than if their surface area were smaller. There is greater water potential in the soil than in the cytoplasm of the root hair cells. As the cells surface membrane of the root hair cell is semi – permeable, osmosis can take place; and water passes from the soil to the root hairs. The next stage in the transpiration stream is water passing to the xylem vessels.

The water either goes through the cortex cells (between the root cells and the xylem vessels) or it bypasses them – going through their cell walls. After this, the water moves up the xylem vessels to the leaves through diffusion: A pressure change between the top and bottom of the vessel. Diffusion takes place because there is a concentration gradient between water in the xylem vessel and the leaf (as water is transpiring out of the leaf). This means that water diffuses up the leaf. There is also a pressure change between the top and bottom of the xylem vessels, due to water loss from the leaves. This reduces the pressure of water at the top of the vessels.

This means water moves up the vessels. The last stage in the transpiration stream is the water moving into the leaves, and then the actual transpiration. First, the water moves into the mesophyll cells from the top of the xylem vessels. Then the water evaporates out of the cells into the spaces between the cells in the leaf. After this, the water leaves the leaf (and the whole plant) by diffusion through stomata.

6

Interfacing with Air

Photosynthesis

Photosynthesis is a process used by plants and other organisms to convert light energy, normally from the sun, into chemical energy that can be later released to fuel the organisms' activities. This chemical energy is stored in carbohydrate molecules, such as sugars, which are synthesized from carbon dioxide and water – hence the name photosynthesis, from the Greek ööò, *phôs*, "light", and óýíèåóéò, *synthesis*, "putting together". In most cases, oxygen is also released as a waste product. Most plants, most algae, and cyanobacteria perform photosynthesis, and such organisms are called photoautotrophs. Photosynthesis maintains atmospheric oxygen levels and supplies all of the organic compounds and most of the energy necessary for life on Earth.

Although photosynthesis is performed differently by different species, the process always begins when energy from light is absorbed by proteins called reaction centres that contain green chlorophyll pigments. In plants, these proteins are held inside organelles called chloroplasts, which are most abundant in leaf cells, while in bacteria they are embedded in the plasma membrane. In these light-dependent reactions, some energy is used to strip electrons from suitable substances such as water, producing oxygen gas. Furthermore, two further compounds are generated: reduced nicotinamide adenine dinucleotide phosphate (NADPH) and adenosine triphosphate (ATP), the "energy currency" of cells.

In plants, algae and cyanobacteria, sugars are produced by a subsequent sequence of light-independent reactions called the Calvin cycle, but some bacteria use different mechanisms, such as the reverse Krebs cycle. In the Calvin cycle, atmospheric carbon dioxide is incorporated into already existing organic carbon compounds, such as ribulose bisphosphate (RuBP). Using the ATP and NADPH produced by the light-dependent reactions, the resulting compounds are then reduced and removed to form further carbohydrates such as glucose.

The first photosynthetic organisms probably evolved early in the evolutionary history of life and most likely used reducing agents such as hydrogen or hydrogen sulfide as sources of electrons, rather than water. Cyanobacteria appeared later, and the excess oxygen they produced contributed to the oxygen catastrophe, which rendered the evolution of complex life possible. Today, the average rate of energy capture by photosynthesis globally is approximately 130 terawatts, which is about six times larger than the current power consumption of human civilization. Photosynthetic organisms also convert around 100–115 thousand million metric tonnes of carbon into biomass per year.

Overview

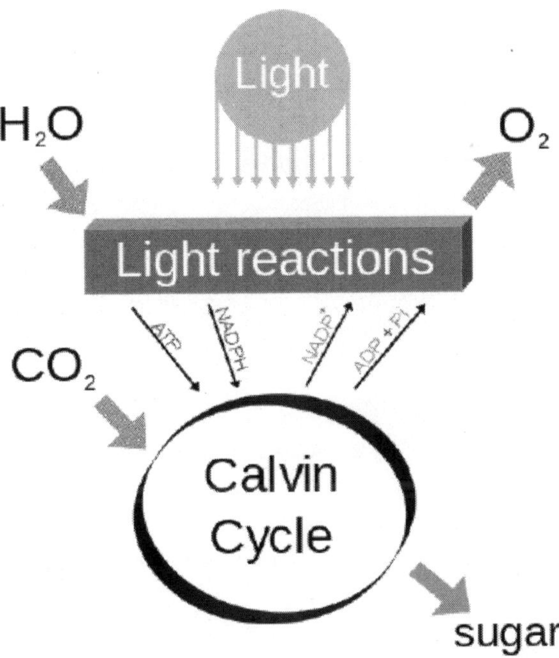

Figure: *Photosynthesis changes sunlight into chemical energy, splits water to liberate O_2, and fixes CO_2 into sugar.*

Photosynthetic organisms are photoautotrophs, which means that they are able to synthesize food directly from carbon dioxide and water using energy from light. However, not all organisms that use light as a source of energy carry out photosynthesis, since *photoheterotrophs* use organic compounds, rather than carbon dioxide, as a source of carbon. In plants, algae and cyanobacteria, photosynthesis releases oxygen. This is called *oxygenic photosynthesis*. Although there are some differences between oxygenic photosynthesis in plants, algae, and cyanobacteria, the overall process is quite similar in these organisms. However, there are some types of bacteria that carry out anoxygenic photosynthesis, which consumes carbon dioxide but does not release oxygen.

Carbon dioxide is converted into sugars in a process called carbon fixation. Carbon fixation is an endothermic redox reaction, so photosynthesis needs to supply both a source of energy to drive this process, and the electrons needed to convert carbon dioxide into a carbohydrate. This addition of the electrons is a reduction reaction. In general outline and in effect, photosynthesis is the opposite of cellular respiration, in which glucose and other compounds are oxidized to produce carbon dioxide and water, and to release exothermic chemical energy to drive the organism's metabolism. However, the two processes take place through a different sequence of chemical reactions and in different cellular compartments.

The general equation for photosynthesis is therefore:

$$2n\ CO_2 + 2n\ DH_2 + \text{photons} \rightarrow 2(CH_2O)_n + 2n\ DO$$

Carbon dioxide + electron donor + light energy → carbohydrate + oxidized electron donor

In *oxygenic* photosynthesis water is the electron donor and, since its hydrolysis releases oxygen, the equation for this process is:

$$2n\ CO_2 + 4n\ H_2O + \text{photons} \rightarrow 2(CH_2O)_n + 2n\ O_2 + 2n\ H_2O$$

carbon dioxide + water + light energy → carbohydrate + oxygen + water

Often 2n water molecules are cancelled on both sides, yielding:

$$2n\ CO_2 + 2n\ H_2O + \text{photons} \rightarrow 2(CH_2O)_n + 2n\ O_2$$

carbon dioxide + water + light energy → carbohydrate + oxygen

Other processes substitute other compounds (such as arsenite) for water in the electron-supply role; for example some microbes use sunlight to oxidize arsenite to arsenate: The equation for this reaction is:

CO_2 + (AsO_3^{3-}) + photons → (AsO_4^{3-}) + CO

carbon dioxide + arsenite + light energy → arsenate + carbon monoxide (used to build other compounds in subsequent reactions)

Photosynthesis occurs in two stages. In the first stage, *light-dependent reactions* or *light reactions* capture the energy of light and use it to make the energy-storage molecules ATP and NADPH. During the second stage, the *light-independent reactions* use these products to capture and reduce carbon dioxide.

Most organisms that utilise photosynthesis to produce oxygen use visible light to do so, although at least three use shortwave infrared or, more specifically, far-red radiation.

Chloroplast

Chloroplasts are organelles, specialised subunits, in plant and algal cells. Their main role is to conduct photosynthesis, where the photosynthetic pigment chlorophyll captures the energy from sunlight, and stores it in the energy storage molecules ATP and NADPH while freeing oxygen from water. They then use the ATP and NADPH to make organic molecules from carbon dioxide in a process known as the Calvin cycle. Chloroplasts carry out a number of other functions, including fatty acid synthesis, much amino acid synthesis, and the immune response in plants.

A chloroplast is one of three types of plastid, characterized by its high concentration of chlorophyll. (The other two types, the leucoplast and the chromoplast, contain little chlorophyll and do not carry out photosynthesis.) Chloroplasts are highly dynamic—they circulate and are moved around within plant cells, and occasionally pinch in two to reproduce. Their behaviour is strongly influenced by environmental factors like light colour and intensity. Chloroplasts, like mitochondria, contain their own DNA, which is thought to be inherited from their ancestor—a photosynthetic cyanobacterium that was engulfed by an early eukaryotic cell. Chloroplasts cannot be made by the plant cell, and must be inherited by each daughter cell during cell division.

With one exception (a member of the genus *Paulinella*), all chloroplasts can probably be traced back to a single endosymbiotic event (the cyanobacterium being engulfed by the eukaryote). Despite this, chloroplasts can be found in an extremely wide set of organisms, some not even directly related to each other—a consequence of many secondary and even tertiary endosymbiotic events.

The word *chloroplast* is derived from the Greek words *chloros*, which means green, and *plastes*, which means "the one who forms".

Chloroplast Lineages and Evolution

Chloroplasts are one of many types of organelles in the plant cell. They are considered to have originated from cyanobacteria through endosymbiosis—when a eukaryotic cell engulfed a photosynthesizing cyanobacterium which remained and became a permanent resident in the cell. Mitochondria are thought to have come from a similar event, where an ærobic prokaryote was engulfed.

This origin of chloroplasts was first suggested by Konstantin Mereschkowski in 1905 after Andreas Schimper observed that chloroplasts closely resemble cyanobacteria in 1883. Chloroplasts are only found in plants and algae.

Cyanobacterial Ancestor

Cyanobacteria are considered the ancestors of chloroplasts. They are sometimes called blue-green algae even though they are prokaryotes. They are a diverse phylum of bacteria capable of carrying out photosynthesis, and are gram-negative, meaning they have two cell membranes.

They also contain a peptidoglycan cell wall, which is thicker than in other gram-negative bacteria, and which is located between their two cell membranes. Like chloroplasts, they have thylakoids inside of them. On the thylakoid membranes are photosynthetic pigments, including chlorophyll *a*. Phycobilins are also common cyanobacterial pigments, usually organised into hemispherical phycobilisomes attached to the outside of the thylakoid membranes (phycobilins are not shared with all chloroplasts though).

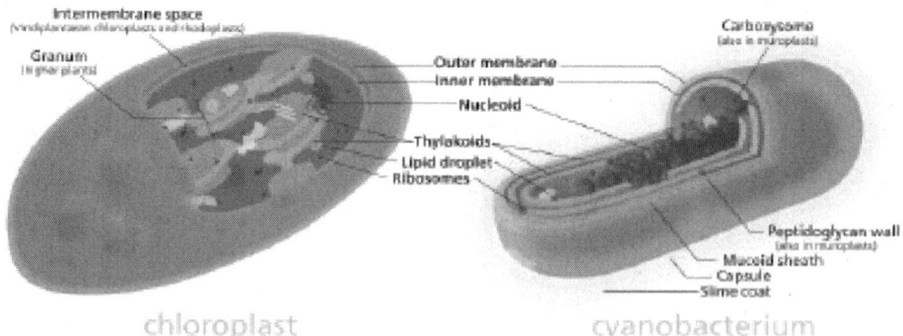

Figure: *Both chloroplasts and cyanobacteria have a double membrane, DNA, ribosomes, and thylakoids. Both the chloroplast and cyanobacterium depicted are idealized versions (the chloroplast is that of a higher plant)—a lot of diversity exists among chloroplasts and cyanobacteria.*

Primary Endosymbiosis

Somewhere around a billion years ago, a free-living cyanobacterium entered an early eukaryotic cell, either as food or an internal parasite, and managed to escape the phagocytic vacuole it was contained in. The two innermost lipid-bilayer membranes that surround all chloroplasts correspond to the outer and inner membranes of the ancestral cyanobacterium's gram negative cell wall, and not the phagosomal membrane from the host, which was probably lost. The new cellular resident quickly became an advantage, providing food for the eukaryotic host, which allowed it to live within it. Over time, the cyanobacterium was assimilated, and many of its genes were lost or transferred to the nucleus of the host. Some of its proteins were then synthesized in the cytoplasm of the host cell, and imported back into the chloroplast.

This event is called *endosymbiosis*, or "cell living inside another cell". The cell living inside the other cell is called the *endosymbiont*; the endosymbiont is found inside the *host cell.*

Chloroplasts are believed to have arisen after mitochondria, since all eukaryotes contain mitochondria, but not all have chloroplasts. This is called *serial endosymbiosis*—an early eukaryote engulfing the mitochondrion ancestor, and some descendants of it then engulfing the chloroplast ancestor, creating a cell with both chloroplasts and mitochondria.

Whether or not chloroplasts came from a single endosymbiotic event, or many independent engulfments across various eukaryotic lineages has been long debated, but it is now generally held that all organisms with chloroplasts either share a single ancestor or obtained their chloroplast from organisms that share a common ancestor that took in a cyanobacterium 600–1600 million years ago.

These chloroplasts, which can be traced back directly to a cyanobacterial ancestor are known as *primary plastids* (*"plastid"* in this context means the almost the same thing as chloroplast). All primary chloroplasts belong to one of three chloroplast lineages—the glaucophyte chloroplast lineage, the rhodophyte, or red algal chloroplast lineage, or the chloroplastidan, or green chloroplast lineage. The second two are the largest, and the green chloroplast lineage is the one that contains the land plants.

Glaucophyta

The alga *Cyanophora*, a glaucophyte, is thought to be one of the first organisms to contain a chloroplast. The glaucophyte chloroplast group is the smallest of the three primary chloroplast lineages, being found in only thirteen species, and is thought to be the one that branched off the earliest. Glaucophytes have chloroplasts which retain a peptidoglycan wall between their double membranes, like their cyanobacterial parent. For this reason, glaucophyte chloroplasts are also known as *muroplasts*. Glaucophyte chloroplasts also contain concentric unstacked thylakoids which surround a carboxysome, an icosahedral structure that glaucophyte chloroplasts and cyanobacteria keep their carbon fixation enzyme rubisco in. The starch they synthesize collects outside the chloroplast. Like cyanobacteria, glaucophyte chloroplast thylakoids are studded with light collecting structures called phycobilisomes. For these reasons, glaucophyte chloroplasts are considered a primitive intermediate between cyanobacteria and the more evolved chloroplasts in red algae and plants.

Rhodophyceæ (Red Algae)

The rhodophyte, or red algal chloroplast group is another large and diverse chloroplast lineage. Rhodophyte chloroplasts are also called *rhodoplasts*, literally "red chloroplasts".

Rhodoplasts have a double membrane with an intermembrane space and phycobilin pigments organised into phycobilisomes on the thylakoid membranes, preventing their thylakoids from stacking. Some contain pyrenoids. Rhodoplasts have chlorophyll *a* and phycobilins for photosynthetic pigments; the phycobilin phycoerytherin is responsible for giving many red algae their distinctive red colour. However, since they also contain the blue-green chlorophyll *a* and other pigments, many are reddish to purple from the combination. The red phycoerytherin pigment is an adaptation to help red algae catch more sunlight in deep water—as such, some red algae that live in shallow water have less phycoerytherin in their rhodoplasts, and can appear more greenish. Rhodoplasts synthesize a form of starch called floridean, which collects into granules outside the rhodoplast, in the cytoplasm of the red alga.

Chloroplastida (Green Algae and Plants)

The chloroplastidan chloroplasts, or green chloroplasts, are another large, highly diverse primary chloroplast lineage. Their host organisms are commonly known as the green algae and land plants. They differ from glaucophyte and red algal chloroplasts in that they have lost their phycobilisomes, and contain chlorophyll *b* instead. Most green chloroplasts are (obviously) green, though some aren't, like some forms of *Hæmatococcus pluvialis*, due to accessory pigments that override the chlorophylls' green colours. Chloroplastidan chloroplasts have lost the peptidoglycan wall between their double membrane, and have replaced it with an intermembrane space. Some plants seem to have kept the genes for the synthesis of the peptidoglycan layer, though they've been repurposed for use in chloroplast division instead.

Most of the chloroplasts depicted in this article are green chloroplasts.

Green algae and plants keep their starch *inside* their chloroplasts, and in plants and some algae, the chloroplast thylakoids are arranged in grana stacks. Some green algal chloroplasts contain a structure called a pyrenoid, which is functionally similar to the glaucophyte carboxysome in that it is where rubisco and CO_2 are concentrated in the chloroplast.

Helicosproidia

The helicosproidia are nonphotosynthetic parasitic green algae that are thought to contain a vestigial chloroplast. Genes from a chloroplast and nuclear genes indicating the presence of a chloroplast have been found in helicosporoidia. even if nobody's seen the chloroplast itself.

Secondary and Tertiary Endosymbiosis

Many other organisms obtained chloroplasts from the primary chloroplast lineages through secondary endosymbiosis—engulfing a red or green alga that contained a chloroplast. These chloroplasts are known as secondary plastids.

While primary chloroplasts have a double membrane from their cyanobacterial ancestor, secondary chloroplasts have additional membranes outside of the original two, as a result of the secondary endosymbiotic event, when a nonphotosynthetic eukaryote engulfed a chloroplast-containing alga but failed to digest it—much like the cyanobacterium at the beginning of this story. The engulfed alga was broken down, leaving only its chloroplast, and sometimes its cell membrane and nucleus, forming a chloroplast with three to four membranes—the two cyanobacterial membranes, sometimes the eaten alga's cell membrane, and the phagosomal vacuole from the host's cell membrane.

The genes in the phagocytosed eukaryote's nucleus are often transferred to the secondary host's nucleus. Cryptomonads and chlorarachniophytes retain the phagocytosed eukaryote's nucleus, an object called a nucleomorph, located between the second and third membranes of the chloroplast. All secondary chloroplasts come from green and red algae—no secondary chloroplasts from glaucophytes have been observed, probably because glaucophytes are relatively rare in nature, making them less likely to have been taken up by another eukaryote.

Green Algal Derived Chloroplasts

Green algae have been taken up by the euglenids, chlorarachniophytes, a lineage of dinoflagellates, and possibly the ancestor of the chromalveolates in three or four separate engulfments. Many green algal derived chloroplasts contain pyrenoids, but unlike chloroplasts in their green algal ancestors, starch collects in granules outside the chloroplast.

Euglenophytes

Euglenophytes are a group of common flagellated protists that contain chloroplasts derived from a green alga. Euglenophyte chloroplasts have three membranes—it is thought that the membrane of the primary endosymbiont was lost, leaving the cyanobacterial membranes, and the secondary host's phagosomal membrane. Euglenophyte chloroplasts have a pyrenoid and thylakoids stacked in groups of three. Starch is stored in the form of paramylon, which is contained in membrane-bound granules in the cytoplasm of the euglenophyte.

Chlorarachniophytes

Chlorarachniophytes are a rare group of organisms that also contain chloroplasts derived from green algae, though their story is more complicated than that of the euglenophytes. The ancestor of chlorarachniophytes is thought to have been a chromalveolate, a eukaryote with a *red* algal derived chloroplast. It is then thought to have lost its first red algal chloroplast, and later engulfed a green alga, giving it its second, green algal derived chloroplast.

Chlorarachniophyte chloroplasts are bounded by four membranes, except near the cell membrane, where the chloroplast membranes fuse into a double membrane. Their thylakoids are arranged in loose stacks of three. Chlorarachniophytes have a form of starch called chrysolaminarin, which they store in the cytoplasm, often collected around the chloroplast pyrenoid, which bulges into the cytoplasm.

Chlorarachniophyte chloroplasts are notable because the green alga they are derived from has not been completely broken down—its nucleus still persists as a nucleomorph found between the second and third chloroplast membranes—the periplastid space, which corresponds to the green alga's cytoplasm.

Early Chromalveolates

Recent research has suggested that the ancestor of the chromalveolates acquired a green algal prasinophyte endosymbiont. The green algal derived chloroplast was lost and replaced with a red algal derived chloroplast, but not before contributing some of its genes to the early chromalveolate's nucleus. The presence of both green algal and red algal genes in chromalveolates probably helps them thrive under fluctuating light conditions.

Red Algal Derived Chloroplasts (Chromalveolate Chloroplasts)

Like green algae, red algae have also been taken up in secondary endosymbiosis, though it is thought that all red algal derived chloroplasts are descended from a single red alga that was engulfed by an early chromalveolate, giving rise to the chromalveolates, some of which, like the ciliates, subsequently lost the chloroplast. This is still debated though.

Pyrenoids and stacked thylakoids are common in chromalveolate chloroplasts, and the outermost membrane of many are continuous with the rough endoplasmic reticulum and studded with ribosomes. They have lost their phycobilisomes and exchanged them for chlorophyll *c*, which isn't found in primary red algal chloroplasts themselves.

Cryptophytes

Cryptophytes, or cryptomonads are a group of algae that contain a red-algal derived chloroplast. Cryptophyte chloroplasts contain a nucleomorph that superficially resembles that of the chlorarachniophytes. Cryptophyte chloroplasts have four membranes, the outermost of which is continuous with the rough endoplasmic reticulum. They synthesize ordinary starch, which is stored in granules found in the periplastid space—outside the original double membrane, in the place that corresponds to the red alga's cytoplasm. Inside cryptophyte chloroplasts is a pyrenoid and thylakoids in stacks of two.

Their chloroplasts do not have phycobilisomes, but they do have phycobilin pigments which they keep in their thylakoid space, rather than anchored on the outside of their thylakoid membranes.

Haptophytes

Haptophytes are similar and closely related to cryptophytes, and are thought to be the first chromalveolates to branch off. Their chloroplasts lack a nucleomorph, their thylakoids are in stacks of three, and they synthesize chrysolaminarin sugar, which they store completely outside of the chloroplast, in the cytoplasm of the haptophyte.

Heterokontophytes (Stramenopiles)

The heterokontophytes, also known as the stramenopiles, are a very large and diverse group of algae that also contain red algal derived chloroplasts. Heterokonts include the diatoms and the brown algae, golden algae, and yellow-green algae.

Heterokont chloroplasts are very similar to haptophyte chloroplasts, containing a pyrenoid, triplet thylakoids, and with some exceptions, having an epiplastid membrane connected to the endoplasmic reticulum. Like haptophytes, heterokontophytes store sugar in chrysolaminarin granules in the cytoplasm. Heterokontophyte chloroplasts contain chlorophyll *a* and with a few exceptions chlorophyll *c*, but also have carotenoids which give them their many

Apicomplexans

Apicomplexans are another group of chromalveolates. Like the helicosproidia, they're parasitic, and have a nonphotosynthetic chloroplast. They were once thought to be related to the helicosproidia, but it is now known that the helicosproida are green algae rather than chromalveolates. The apicomplexans include *Plasmodium*, the malaria parasite. Many apicomplexans keep a vestigial red algal derived chloroplast called an apicoplast, which they inherited from their ancestors. Other apicomplexans like *Cryptosporidium* have lost the chloroplast completely. Apicomplexans store their energy in amylopectin starch granules that are located in their cytoplasm, even though they are nonphotosynthetic.

Apicoplasts have lost all photosynthetic function, and contain no photosynthetic pigments or true thylakoids. They are bounded by four membranes, but the membranes are not connected to the endoplasmic reticulum.

The fact that apicomplexans still keep their nonphotosynthetic chloroplast around demonstrates how the chloroplast carries out important functions other than photosynthesis. Plant chloroplasts provide plant cells with many important things besides sugar, and apicoplasts are no different—they synthesize fatty acids, isopentenyl pyrophosphate, iron-sulfur clusters, and carry out part of the heme pathway. This makes the apicoplast an attractive target for drugs to cure apicomplexan-related diseases.

The most important apicoplast function is isopentenyl pyrophosphate synthesis—in fact, apicomplexans die when something interferes with this apicoplast function, and when apicomplexans are grown in an isopentenyl pyrophosphate-rich medium, they dump the organelle.

Dinophytes

The dinoflagellates are yet another very large and diverse group of protists, around half of which are (at least partially) photosynthetic.

Most dinophyte chloroplasts are secondary red algal derived chloroplasts, like other chromalveolate chloroplasts. Many other dinophytes have lost the chloroplast (becoming the nonphotosynthetic kind of dinoflagellate), or replaced it though *tertiary* endosymbiosis—the engulfment of another chromalveolate containing a red algal derived chloroplast. Others replaced their original chloroplast with a green algal derived one.

Most dinophyte chloroplasts contain at least the photosynthetic pigments chlorophyll *a*, chlorophyll c_2, *beta*-carotene, and at least one dinophyte-unique xanthophyll (peridinin, dinoxanthin, or diadinoxanthin), giving many a golden-brown colour. All dinophytes store starch in their cytoplasm, and most have chloroplasts with thylakoids arranged in stacks of three.

Peridinin-Containing Dinophyte Chloroplast

The most common dinophyte chloroplast is the peridinin-type chloroplast, characterized by the carotenoid pigment peridinin in their chloroplasts, along with chlorophyll *a* and chlorophyll c_2. Peridinin is not found in any other group of chloroplasts. The peridinin chloroplast is bounded by three membranes (occasionally two), having lost the red algal endosymbiont's original cell membrane.

The outermost membrane is not connected to the endoplasmic reticulum.

They contain a pyrenoid, and have triplet-stacked thylakoids. Starch is found outside the chloroplast An important feature of these chloroplasts is that their chloroplast DNA is highly reduced and fragmented into many small circles. Most of the genome has migrated to the nucleus, and only critical photosynthesis-related genes remain in the chloroplast.

The peridinin chloroplast is thought to be the dinophytes' "original" chloroplast, which has been lost, reduced, replaced, or has company in several other dinophyte lineages.

Fucoxanthin-Containing Dinophyte Chloroplasts (Haptophyte Endosymbionts)

The fucoxanthin dinophyte lineages (including *Karlodinium* and *Karenia*) lost their original red algal derived chloroplast, and replaced it with a new chloroplast derived from a haptophyte endosymbiont. *Karlodinium* and *Karenia* probably took up different heterokontophytes. Because the haptophyte chloroplast has four membranes, tertiary endosymbiosis would be expected to create a six membraned chloroplast, adding the haptophyte's cell membrane and the dinophyte's phagosomal vacuole. However, the haptophyte was heavily reduced, stripped of a few membranes and its nucleus, leaving only its chloroplast (with its original double membrane), and possibly one or two additional membranes around it.

Fucoxanthin-containing chloroplasts are characterized by having the pigment fucoxanthin (actually 192 -hexanoyloxy-fucoxanthin and/ or 192 -butanoyloxy-fucoxanthin) and no peridinin. Fucoxanthin is also found in haptophyte chloroplasts, providing evidence of ancestry.

Cryptophyte Derived Dinophyte Chloroplast

Members of the genus *Dinophysis* have a phycobilin-containing chloroplast taken from a cryptophyte. However, the cryptophyte is not an endosymbiont—only the chloroplast seems to have been taken, and the chloroplast has been stripped of its nucleomorph and outermost two membranes, leaving just a two-membraned chloroplast. Cryptophyte chloroplasts require their nucleomorph to maintain themselves, and *Dinophysis* species grown in cell culture alone cannot survive, so it is possible (but not confirmed) that the *Dinophysis* chloroplast is a kleptoplast—if so, *Dinophysis* chloroplasts wear out and *Dinophysis* species must continually engulf cryptophytes to obtain new chloroplasts to replace the old ones.

Diatom Derived Dinophyte Chloroplasts

Some dinophytes, like *Kryptoperidinium* and *Durinskia* have a diatom (heterokontophyte) derived chloroplast. These chloroplasts are bounded by up to *five* membranes, (depending on whether you count the entire diatom endosymbiont as the chloroplast, or just the red algal derived chloroplast inside it). The diatom endosymbiont has been reduced relatively little—it still retains its original mitochondria, and has endoplasmic reticulum, ribosomes, a nucleus, and of course, red algal derived chloroplasts—practically a complete cell, all inside the host's endoplasmic reticulum lumen. However the diatom endosymbiont can't store its own food—its starch is found in granules in the dinophyte host's cytoplasm instead. The diatom endosymbiont's nucleus is present, but it probably can't be called a nucleomorph because it shows no sign of genome reduction, and might have even been *expanded*. Diatoms have been engulfed by dinoflagellates at least three times.

The diatom endosymbiont is bounded by a single membrane, inside it are chloroplasts with four membranes. Like the diatom endosymbiont's diatom ancestor, the chloroplasts have triplet thylakoids and pyrenoids.

In some of these genera, the diatom endosymbiont's chloroplasts aren't the only chloroplasts in the dinophyte. The original three-membraned peridinin chloroplast is still around, converted to an eyespot.

Prasinophyte (Green Algal) Derived Dinophyte Chloroplast

Lepidodinium viride and its close relatives are dinophytes that lost their original peridinin chloroplast and replaced it with a green algal derived chloroplast (more specifically, a prasinophyte). *Lepidodinium* is the only dinophyte that has a chloroplast that's not from the rhodoplast lineage. The chloroplast is surrounded by two membranes and has no nucleomorph—all the nucleomorph genes have been transferred to the dinophyte nucleus. The endosymbiotic event that led to this chloroplast was serial secondary endosymbiosis rather than tertiary endosymbiosis—the endosymbiont was a green alga containing a primary chloroplast (making a secondary chloroplast).

Chromatophores

While most chloroplasts originate from that first set of endosymbiotic events, *Paulinella chromatophora* is an exception, which has acquired a photosynthetic cyanobacterial endosymbiont more recently. It is not clear whether that symbiont is closely related to the ancestral chloroplast of other eukaryotes. Being in the early stages of endosymbiosis, *Paulinella chromatophora* can offer some insights into how chloroplasts evolved. *Paulinella* cells contain one or two sausage shaped blue-green photosynthesizing structures called chromatophores, descended from the cyanobacterium *Synechococcus*. Chromatophores cannot survive outside their host. Chromatophore DNA is about a million base pairs long, containing around 850 protein encoding genes—far less than the three million base pair *Synechococcus* genome, but much larger than the approximately 150,000 base pair genome of the more assimilated chloroplast. Chromatophores have transferred much less of their DNA to the nucleus of their host. About

0.3–0.8% of the nuclear DNA in *Paulinella* is from the chromatophore, compared with 11–14% from the chloroplast in plants.

Kleptoplastidy

In some groups of mixotrophic protists, like some dinoflagellates, chloroplasts are separated from a captured alga or diatom and used temporarily. These klepto chloroplasts may only have a lifetime of a few days and are then replaced.

Chloroplast DNA

Chloroplasts have their own DNA, often abbreviated as ctDNA, or cpDNA. It is also known as the plastome. Its existence was first proved in 1962, and first sequenced in 1986—when two Japanese research teams sequenced the chloroplast DNA of liverwort and tobacco. Since then, hundreds of chloroplast DNAs from various species have been sequenced, but they're mostly those of land plants and green algae—glaucophytes, red algae, and other algal groups are extremely under-represented, potentially introducing some bias in views of "typical" chloroplast DNA structure and content.

Inverted Repeats

Many chloroplast DNAs contain two *inverted repeats,* which separate a long single copy section (LSC) from a short single copy section (SSC). While a given pair of inverted repeats are rarely completely identical, they are always very similar to each other, apparently resulting from concerted evolution. The inverted repeats vary wildly in length, ranging from 4,000 to 25,000 base pairs long each and containing as few as four or as many as over 150 genes. Inverted repeats in plants tend to be at the upper end of this range, each being 20,000–25,000 base pairs long.

The inverted repeat regions are highly conserved among land plants, and accumulate few mutations. Similar inverted repeats exist in the genomes of cyanobacteria and the other two chloroplast lineages (glaucophyta and rhodophyceæ), suggesting that they predate the chloroplast, though some chloroplast DNAs have since lost or flipped the inverted repeats (making them direct repeats). It is possible that the inverted repeats help stabilize the rest of the chloroplast genome, as chloroplast DNAs which have lost some of the inverted repeat segments tend to get rearranged more.

Nucleoids

New chloroplasts may contain up to 100 copies of their DNA, though the number of chloroplast DNA copies decreases to about 15–

20 as the chloroplasts age. They are usually packed into nucleoids which can contain several identical chloroplast DNA rings. Many nucleoids can be found in each chloroplast. In primitive red algae, the chloroplast DNA nucleoids are clustered in the centre of the chloroplast, while in green plants and green algae, the nucleoids are dispersed throughout the stroma.

Though chloroplast DNA is not associated with true histones, in red algae, similar proteins that tightly pack each chloroplast DNA ring into a nucleoid have been found.

DNA Replication

Gene Content and Protein Synthesis

The chloroplast genome most commonly includes around 100 genes which code for a variety of things, mostly to do with the protein pipeline and photosynthesis. As in prokaryotes, genes in chloroplast DNA are organised into operons. Interestingly though, unlike prokaryotic DNA molecules, chloroplast DNA molecules contain introns (plant mitochondrial DNAs do too, but not human mtDNAs).

Among land plants, the contents of the chloroplast genome are fairly similar.

Chloroplast Genome Reduction and Gene Transfer

Over time, many parts of the chloroplast genome were transferred to the nuclear genome of the host, a process called *endosymbiotic gene transfer*. As a result, the chloroplast genome is heavily reduced compared to that of free-living cyanobacteria. Chloroplasts may contain 60–100 genes whereas cyanobacteria often have more than 1500 genes in their genome.

Endosymbiotic gene transfer is how we know about the lost chloroplasts in many chromalveolate lineages. Even if a chloroplast is eventually lost, the genes it donated to the former host's nucleus persist, providing evidence for the lost chloroplast's existence. For example, while diatoms (a heterokontophyte) now have a red algal derived chloroplast, the presence of many green algal genes in the diatom nucleus provide evidence that the diatom ancestor (probably the ancestor of all chromalveolates too) had a green algal derived chloroplast at some point, which was subsequently replaced by the red chloroplast.

In land plants, some 11–14% of the DNA in their nuclei can be traced back to the chloroplast, up to 18% in *Arabidopsis*, corresponding to about 4,500 protein-coding genes. There have been a few recent transfers of genes from the chloroplast DNA to the nuclear genome in land plants. Of the approximately three-thousand proteins found in chloroplasts, some 95% of them are encoded by nuclear genes. Many of the chloroplast's protein complexes consist of subunits from both the chloroplast genome and the host's nuclear genome. As a result, protein synthesis must be coordinated between the chloroplast and the nucleus. The chloroplast is mostly under nuclear control, though chloroplasts can also give out signals regulating gene expression in the nucleus, called *retrograde signalling.*

Protein Synthesis

Protein synthesis within chloroplasts relies on two RNA polymerases. One is coded by the chloroplast DNA, the other is of nuclear origin. The two RNA polymerases may recognise and bind to different kinds of promoters within the chloroplast genome. The ribosomes in chloroplasts are similar to bacterial ribosomes.

Protein Targeting and Import

The movement of so many chloroplast genes to the nucleus means that lots of chloroplast proteins that were supposed to be translated in the chloroplast are now synthesized in the cytoplasm. This means that these proteins must be directed back to the chloroplast, and imported through at least two chloroplast membranes.

Curiously, around half of the protein products of transferred genes aren't even targeted back to the chloroplast. Many became exaptations, taking on new functions like participating in cell division, protein routing, and even disease resistance. A few chloroplast genes found new homes in the mitochondrial genome—most became nonfunctional pseudogenes, though a few tRNA genes still work in the mitochondrion. Some transferred chloroplast DNA protein products get directed to the secretory pathway (though it should be noted that many secondary plastids are bounded by an outermost membrane derived from the host's cell membrane, and therefore topologically outside of the cell, because to reach the chloroplast from the cytosol, you have to cross the cell membrane, just like if you were headed for the extracellular space. In those cases, chloroplast-targeted proteins do initially travel along the secretory pathway).

Because the cell acquiring a chloroplast already had mitochondria (and peroxisomes, and a cell membrane for secretion), the new chloroplast host had to develop a unique protein targeting system to avoid having chloroplast proteins being sent to the wrong organelle, and the wrong proteins being sent to the chloroplast.

The two ends of a polypeptide are called the N-terminus, or *amino end*, and the C-terminus, or *carboxyl end.* This polypeptide has four amino acids linked together. At the left is the N-terminus, with its amino (H_2N) group in green. The blue C-terminus, with its carboxyl group (CO_2H) is at the right.

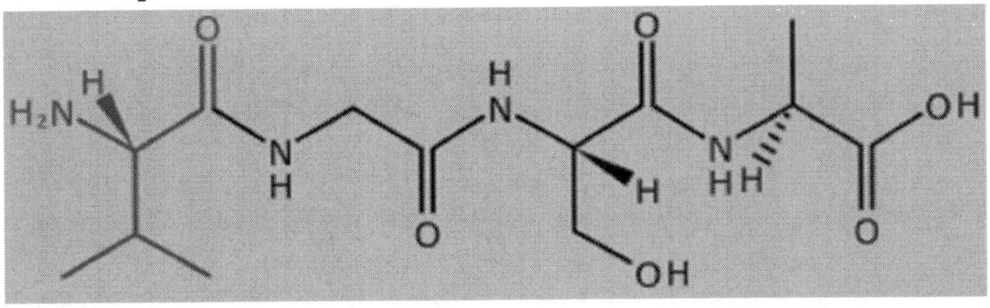

In most, but not all cases, nuclear-encoded chloroplast proteins are translated with a *cleavable transit peptide* that's added to the N-terminus of the protein precursor. Sometimes the transit sequence is found on the C-terminus of the protein, or within the functional part of the protein.

Transport Proteins and MembraneTranslocons

After a chloroplast polypeptide is synthesized on a ribosome in the cytosol, an enzyme specific to chloroplast proteins phosphorylates, or adds a phosphate group to many (but not all) of them in their transit sequences. Phosphorylation helps many proteins bind the polypeptide, keeping it from folding prematurely. This is important because it prevents chloroplast proteins from assuming their active form and carrying out their chloroplast functions in the wrong place—the cytosol. At the same time, they have to keep just enough shape so that they can be recognised by the chloroplast. These proteins also help the polypeptide get imported into the chloroplast.

From here, chloroplast proteins bound for the stroma must pass through two protein complexes—the TOC complex, or *translocon on the outer chloroplast membrane,* and the TIC translocon, or *translocon on the inner chloroplast membrane translocon.* Chloroplast polypeptide chains probably often travel through the two complexes at the same time, but the TIC complex can also retrieve preproteins lost in the intermembrane space.

Structure

In land plants, chloroplasts are generally lens-shaped, 5–8 ìm in diameter and 1–3 ìm thick. Greater diversity in chloroplast shapes exists among the algae, which often contain a single chloroplast that can be shaped like a net (e.g., *Oedogonium*), a cup (e.g., *Chlamydomonas*), a ribbon-like spiral around the edges of the cell (e.g., *Spirogyra*), or slightly twisted bands at the cell edges (e.g., *Sirogonium*). Some algae have two chloroplasts in each cell; they are star-shaped in *Zygnema*, or may follow the shape of half the cell in order Desmidiales. In some algae, the chloroplast takes up most of the cell, with pockets for the nucleus and other organelles (for example some species of *Chlorella* have a cup-shaped chloroplast that occupies much of the cell).

All chloroplasts have at least three membrane systems—the outer chloroplast membrane, the inner chloroplast membrane, and the thylakoid system. Chloroplasts that are the product of secondary endosymbiosis may have additional membranes surrounding these three. Inside the outer and inner chloroplast membranes is the chloroplast stroma, a semi-gel-like fluid that makes up much of a chloroplast's volume, and in which the thylakoid system floats.

There are some common misconceptions about the outer and inner chloroplast membranes. The fact that chloroplasts are surrounded by a double membrane is often cited as evidence that they are the descendants of endosymbiotic cyanobacteria. This is often interpreted as meaning the outer chloroplast membrane is the product of the host's cell membrane infolding to form a vesicle to surround the ancestral cyanobacterium—which is not true—both chloroplast membranes are homologous to the cyanobacterium's original double membranes.

The chloroplast double membrane is also often compared to the mitochondrial double membrane. This is not a valid comparison—the inner mitochondria membrane is used to run proton pumps and carry out oxidative phosphorylation across to generate ATP energy. The only chloroplast structure that can considered analogous to it is the internal thylakoid system. Even so, in terms of "in-out", the direction of chloroplast H^+ ion flow is in the opposite direction compared to oxidative phosphorylation in mitochondria. In addition, in terms of function, the inner chloroplast membrane, which regulates metabolite passage and synthesizes some materials, has no counterpart in the mitochondrion.

Outer Chloroplast Membrane

The outer chloroplast membrane is a semi-porous membrane that small molecules and ions can easily diffuse across. However, it is not permeable to larger proteins, so chloroplast polypeptides being synthesized in the cell cytoplasm must be transported across the outer chloroplast membrane by the TOC complex, or *translocon on the outer chloroplast* membrane.

The chloroplast membranes sometimes protrude out into the cytoplasm, forming a stromule, or stroma-containing tubule. Stromules are very rare in chloroplasts, and are much more common in other plastids like chromoplasts and amyloplasts in petals and roots, respectively. They may exist to increase the chloroplast's surface area for cross-membrane transport, because they are often branched and tangled with the endoplasmic reticulum. They were once thought to connect chloroplasts allowing them to exchange proteins, however recent research strongly refutes this idea. Observed stromules are probably just oddly shaped chloroplasts with constricted regions or dividing chloroplasts.

Intermembrane Space and Peptidoglycan Wall

Usually, a thin intermembrane space about 10–20 nanometres thick exists between the outer and inner chloroplast membranes.

Glaucophyte algal chloroplasts have a peptidoglycan layer between the chloroplast membranes. It corresponds to the peptidoglycan cell wall of their cyanobacterial ancestors, which is located between their two cell membranes. These chloroplasts are called muroplasts (from Latin *"mura"*, meaning "wall"). Other chloroplasts have lost the cyanobacterial wall, leaving an intermembrane space between the two chloroplast envelope membranes.

Inner Chloroplast Membrane

The inner chloroplast membrane borders the stroma and regulates passage of materials in and out of the chloroplast. After passing through the TOC complex in the outer chloroplast membrane, polypeptides must pass through the TIC complex *(translocon on the inner chloroplast membrane)* which is located in the inner chloroplast membrane.

In addition to regulating the passage of materials, the inner chloroplast membrane is where fatty acids, lipids, and carotenoids are synthesized.

Peripheral Reticulum

Some chloroplasts contain a structure called the chloroplast peripheral reticulum. It is often found in the chloroplasts of C_4 plants, though it has also been found in some C_3 angiosperms, and even some gymnosperms. The chloroplast peripheral reticulum consists of a maze of membranous tubes and vesicles continuous with the inner chloroplast membrane that extends into the internal stromal fluid of the chloroplast. Its purpose is thought to be to increase the chloroplast's surface area for cross-membrane transport between its stroma and the cell cytoplasm. The small vesicles sometimes observed may serve as transport vesicles to shuttle stuff between the thylakoids and intermembrane space.

Stroma

The protein-rich, alkaline, aqueous fluid within the inner chloroplast membrane and outside of the thylakoid space is called the stroma, which corresponds to the cytosol of the original cyanobacterium. Nucleoids of chloroplast DNA, chloroplast ribosomes, the thylakoid system with plastoglobuli, starch granules, and many proteins can be found floating around in it. The Calvin cycle, which fixes CO_2 into sugar takes place in the stroma.

Chloroplast Ribosomes

Chloroplasts have their own ribosomes, which they use to synthesize a small fraction of their proteins. Chloroplast ribosomes are about two-thirds the size of cytoplasmic ribosomes (around 17 nm vs 25 nm). They take mRNAs transcribed from the chloroplast DNA and translate them into protein. While similar to bacterial ribosomes, chloroplast translation is more complex than in bacteria, so chloroplast ribosomes include some chloroplast-unique features.

Plastoglobuli

Plastoglobuli (singular *plastoglobulus*, sometimes spelled *plastoglobule(s)*), are spherical bubbles of lipids and proteins about 45–60 nanometres across. They are surrounded by a lipid monolayer. Plastoglobuli are found in all chloroplasts, but become more common when the chloroplast is under oxidative stress, or when it ages and transitions into a gerontoplast. Plastoglobuli also exhibit a greater size variation under these conditions. They are also common in etioplasts, but decrease in number as the etioplasts mature into chloroplasts.

Plastoglubuli contain both structural proteins and enzymes involved in lipid synthesis and metabolism. They contain many types of lipids including plastoquinone, vitamin E, carotenoids and chlorophylls.

Plastoglobuli were once thought to be free-floating in the stroma, but it is now thought that they are permanently attached either to a thylakoid or to another plastoglobulus attached to a thylakoid, a configuration that allows a plastoglobulus to exchange its contents with the thylakoid network. In normal green chloroplasts, the vast majority of plastoglobuli occur singularly, attached directly to their parent thylakoid. In old or stressed chloroplasts, plastoglobuli tend to occur in linked groups or chains, still always anchored to a thylakoid.

Plastoglobuli form when a bubble appears between the layers of the lipid bilayer of the thylakoid membrane, or bud from existing plastoglubuli—though they never detach and float off into the stroma. Practically all plastoglobuli form on or near the highly curved edges of the thylakoid disks or sheets. They are also more common on stromal thylakoids than on granal ones.

Starch Granules

Starch granules are very common in chloroplasts, typically taking up 15% of the organelle's volume, though in some other plastids like amyloplasts, they can be big enough to distort the shape of the organelle. Starch granules are simply accumulations of starch in the stroma, and are not bounded by a membrane.

Starch granules appear and grow throughout the day, as the chloroplast synthesizes sugars, and are consumed at night to fuel respiration and continue sugar export into the phloem, though in mature chloroplasts, it is rare for a starch granule to be completely consumed or for a new granule to accumulate.

Starch granules vary in composition and location across different chloroplast lineages. In red algae, starch granules are found in the cytoplasm rather than in the chloroplast. In C_4 plants, mesophyll chloroplasts, which do not synthesize sugars, lack starch granules.

Rubisco

The chloroplast stroma contains many proteins, though the most common and important is Rubisco, which is probably also the most abundant protein on the planet. Rubisco is the enzyme that fixes CO_2 into sugar molecules. In C_3 plants, rubisco is abundant in all chloroplasts, though in C_4 plants, it is confined to the bundle sheath chloroplasts, where the Calvin cycle is carried out in C_4 plants.

Pyrenoids

The chloroplasts of some hornworts and algae contain structures called pyrenoids. They are not found in higher plants. Pyrenoids are roughly spherical and highly refractive bodies which are a site of starch accumulation in plants that contain them. They consist of a matrix opaque to electrons, surrounded by two hemispherical starch plates. The starch is accumulated as the pyrenoids mature. In algae with carbon concentrating mechanisms, the enzyme rubisco is found in the pyrenoids. Starch can also accumulate around the pyrenoids when CO_2 is scarce. Pyrenoids can divide to form new pyrenoids, or be produced "de novo".

Thylakoid System

Suspended within the chloroplast stroma is the thylakoid system, a highly dynamic collection of membranous sacks called thylakoids where chlorophyll is found and the light reactions of photosynthesis happen. In most vascular plant chloroplasts, the thylakoids are arranged in stacks called grana, though in certain C_4 plant chloroplasts and some algal chloroplasts, the thylakoids are free floating.

Granal Structure

Using a light microscope, it is just barely possible to see tiny green granules—which were named grana. With electron microscopy, it became possible to see the thylakoid system in more detail, revealing it to consist of stacks of flat thylakoids which made up the grana, and long interconnecting stromal thylakoids which linked different grana. In the transmission electron microscope, thylakoid membranes appear as alternating light-and-dark bands, 8.5 nanometres thick.

For a long time, the three-dimensional structure of the thylakoid system has been unknown or disputed. One model has the granum as a stack of thylakoids linked by helical stromal thylakoids; the other has the granum as a single folded thylakoid connected in a "hub and spoke" way to other grana by stromal thylakoids. While the thylakoid system is still commonly depicted according to the folded thylakoid model, it was determined in 2011 that the stacked and helical thylakoids model is correct.

In the helical thylakoid model, grana consist of a stack of flattened circular granal thylakoids that resemble pancakes. Each granum can contain anywhere from two to a hundred thylakoids, though grana with 10–20 thylakoids are most common. Wrapped around the grana are helicoid stromal thylakoids, also known as frets or lamellar thylakoids. The helices ascend at an angle of 20–25°, connecting to each granal thylakoid at a bridge-like slit junction. The helicoids may extend as large sheets that link multiple grana, or narrow to tube-like bridges between grana. While different parts of the thylakoid system contain different membrane proteins, the thylakoid membranes are continuous and the thylakoid space they enclose form a single continuous labyrinth.

Thylakoids

Thylakoids (sometimes spelled *thylakoïds*), are small interconnected sacks which contain the membranes that the light reactions of photosynthesis take place on. The word *thylakoid* comes from the Greek word *thylakos* which means "sack".

Embedded in the thylakoid membranes are important protein complexes which carry out the light reactions of photosynthesis. Photosystem II and photosystem I contain light-harvesting complexes with chlorophyll and carotenoids that absorb light energy and use it to energize electrons. Molecules in the thylakoid membrane use the energized electrons to pump hydrogen ions into the thylakoid space, decreasing the pH and turning it acidic. ATP synthase is a large protein complex that harnesses the concentration gradient of the hydrogen ions in the thylakoid space to generate ATP energy as the hydrogen ions flow back out into the stroma—much like a dam turbine.

There are two types of thylakoids—granal thylakoids, which are arranged in grana, and stromal thylakoids, which are in contact with the stroma. Granal thylakoids are pancake-shaped circular disks about 300–600 nanometres in diameter. Stromal thylakoids are helicoid sheets that spiral around grana. The flat tops and bottoms of granal thylakoids contain only the relatively flat photosystem II protein complex. This allows them to stack tightly, forming grana with many layers of tightly appressed membrane, called granal membrane, increasing stability and surface area for light capture.

In contrast, photosystem I and ATP synthase are large protein complexes which jut out into the stroma. They can't fit in the appressed granal membranes, and so are found in the stromal thylakoid membrane—the edges of the granal thylakoid disks and the stromal thylakoids. These large protein complexes may act as spacers between the sheets of stromal thylakoids.

The number of thylakoids and the total thylakoid area of a chloroplast is influenced by light exposure. Shaded chloroplasts contain larger and more grana with more thylakoid membrane area than chloroplasts exposed to bright light, which have smaller and fewer grana and less thylakoid area. Thylakoid extent can change within minutes of light exposure or removal.

Pigments and Chloroplast Colours

Inside the photosystems embedded in chloroplast thylakoid membranes are various photosynthetic pigments, which absorb and transfer light energy. The types of pigments found are different in various groups of chloroplasts, and are responsible for a wide variety of chloroplast colourations.

Chlorophylls

Chlorophyll *a* is found in all chloroplasts, as well as their cyanobacterial ancestors. Chlorophyll *a* is a blue-green pigment partially responsible for giving most cyanobacteria and chloroplasts their colour. Other forms of chlorophyll exist, such as the accessory pigments chlorophyll *b*, chlorophyll *c*, chlorophyll *d*, and chlorophyll *f*.

Chlorophyll *b* is an olive green pigment found only in the chloroplasts of plants, green algae, any secondary chloroplasts obtained through the secondary endosymbiosis of a green alga, and a few cyanobacteria. It is the chlorophylls *a* and *b* together that make most plant and green algal chloroplasts green.

Chlorophyll *c* is mainly found in secondary endosymbiotic chloroplasts that originated from a red alga, although it is not found in chloroplasts of red algae themselves. Chlorophyll *c* is also found in some green algae and cyanobacteria.

Chlorophylls *d* and *f* are pigments found only in some cyanobacteria.

Carotenoids

In addition to chlorophylls, another group of yellow–orange pigments called carotenoids are also found in the photosystems. There are about thirty photosynthetic carotenoids. They help transfer and dissipate excess energy, and their bright colours sometimes override the chlorophyll green, like during the fall, when the leaves of some land plants change colour. â-carotene is a bright red-orange carotenoid found in nearly all chloroplasts, like chlorophyll *a*. Xanthophylls, especially the orange-red zeaxanthin, are also common. Many other forms of carotenoids exist that are only found in certain groups of chloroplasts.

Phycobilins

Phycobilins are a third group of pigments found in cyanobacteria, and glaucophyte, red algal, and cryptophyte chloroplasts. Phycobilins come in all colours, though phycoerytherin is one of the pigments that makes many red algae red. Phycobilins often organise into relatively large protein complexes about 40 nanometres across called phycobilisomes. Like photosystem I and ATP synthase, phycobilisomes jut into the stroma, preventing thylakoid stacking in red algal chloroplasts. Cryptophyte chloroplasts and some cyanobacteria don't have their phycobilin pigments organised into phycobilisomes, and keep them in their thylakoid space instead.

Specialised Chloroplasts in C_4 Plants

To fix carbon dioxide into sugar molecules in the process of photosynthesis, chloroplasts use an enzyme called rubisco. Rubisco has a problem—it has trouble distinguishing between carbon dioxide and oxygen, so at high oxygen concentrations, rubisco starts accidentally adding oxygen to sugar precursors. This has the end result of ATP energy being wasted and CO_2 being released, all with no sugar being produced. This is a big problem, since O_2 is produced by the initial light reactions of photosynthesis, causing issues down the line in the Calvin cycle which uses rubisco.

C_4 plants evolved a way to solve this—by spatially separating the light reactions and the Calvin cycle. The light reactions, which store light energy in ATP and NADPH, are done in the mesophyll cells of a C_4 leaf. The Calvin cycle, which uses the stored energy to make sugar using rubisco, is done in the bundle sheath cells, a layer of cells surrounding a vein in a leaf.

As a result, chloroplasts in C_4 mesophyll cells and bundle sheath cells are specialised for each stage of photosynthesis. In mesophyll cells, chloroplasts are specialised for the light reactions, so they lack rubisco, and have normal grana and thylakoids, which they use to make ATP and NADPH, as well as oxygen. They store CO_2 in a four-carbon compound, which is why the process is called *C_4 photosynthesis*. The four-carbon compound is then transported to the bundle sheath chloroplasts, where it drops off CO_2 and returns to the mesophyll. Bundle sheath chloroplasts do not carry out the light reactions, preventing oxygen from building up in them and disrupting rubisco activity. Because of this, they lack thylakoids organised into grana stacks—though bundle sheath chloroplasts still have free-floating thylakoids in the stroma where they still carry out cyclic electron flow, a light-driven method of synthesizing ATP to power the Calvin cycle without generating oxygen. They lack photosystem II, and only have photosystem I—the only protein complex needed for cyclic electron flow. Because the job of bundle sheath chloroplasts is to carry out the Calvin cycle and make sugar, they often contain large starch grains.

Both types of chloroplast contain large amounts of chloroplast peripheral reticulum, which they use to get more surface area to transport stuff in and out of them. Mesophyll chloroplasts have a little more peripheral reticulum than bundle sheath chloroplasts.

Location

Distribution in a Plant: Not all cells in a multicellular plant contain chloroplasts. All green parts of a plant contain chloroplasts—the chloroplasts, or more specifically, the chlorophyll in them are what make the photosynthetic parts of a plant green. The plant cells which contain chloroplasts are usually parenchyma cells, though chloroplasts can also be found in collenchyma tissue. A plant cell which contains chloroplasts is known as a chlorenchyma cell. A typical chlorenchyma cell of a land plant contains about 10 to 100 chloroplasts.

In some plants such as cacti, chloroplasts are found in the stems, though in most plants, chloroplasts are concentrated in the leaves. One square millimetre of leaf tissue can contain half a million chloroplasts. Within a leaf, chloroplasts are mainly found in the mesophyll layers of a leaf, and the guard cells of stomata. Palisade mesophyll cells can contain 30–70 chloroplasts per cell, while stomatal guard cells contain only around 8–15 per cell, as well as much less chlorophyll. Chloroplasts can also be found in the bundle sheath cells of a leaf, especially in C_4 plants, which carry out the Calvin cycle in their bundle sheath cells. They are often absent from the epidermis of a leaf.

Cellular Location

Chloroplast Movement: The chloroplasts of plant and algal cells can orient themselves to best suit the available light. In low-light conditions, they will spread out in a sheet—maximizing the surface area to absorb light. Under intense light, they will seek shelter by aligning in vertical columns along the plant cell's cell wall or turning sideways so that light strikes them edge-on. This reduces exposure and protects them from photooxidative damage. This ability to distribute chloroplasts so that they can take shelter behind each other or spread out may be the reason why land plants evolved to have many small chloroplasts instead of a few big ones. Chloroplast movement is considered one of the most closely regulated stimulus-response systems that can be found in plants. Mitochondria have also been observed to follow chloroplasts as they move.

In higher plants, chloroplast movement is run by phototropins, blue light photoreceptors also responsible for plant phototropism. In some algae, mosses, ferns, and flowering plants, chloroplast movement is influenced by red light in addition to blue light, though very long red wavelengths inhibit movement rather than speeding it up. Blue light generally causes chloroplasts to seek shelter, while red light draws them out to maximize light absorption.

Studies of *Vallisneria gigantea,* an aquatic flowering plant, have shown that chloroplasts can get moving within five minutes of light exposure, though they don't initially show any net directionality. They may move along microfilament tracks, and the fact that the microfilament mesh changes shape to form a honeycomb structure surrounding the chloroplasts after they have moved suggests that microfilaments may help to anchor chloroplasts in place.

Function and Chemistry

Guard Cell Chloroplasts: Unlike most epidermal cells, the guard cells of plant stomata contain relatively well developed chloroplasts. However, exactly what they do is controversial.

Plant Innate Immunity

Plants lack specialised immune cells—all plant cells participate in the plant immune response. Chloroplasts, along with the nucleus, cell membrane, and endoplasmic reticulum, are key players in pathogen defence. Due to its role in a plant cell's immune response, pathogens frequently target the chloroplast.

Plants have two main immune responses—the hypersensitive response, in which infected cells seal themselves off and undergo programmed cell death, and systemic acquired resistance, where infected cells release signals warning the rest of the plant of a pathogen's presence. Chloroplasts stimulate both responses by purposely damaging their photosynthetic system, producing reactive oxygen species. High levels of reactive oxygen species will cause the hypersensitive response.

The reactive oxygen species also directly kill any pathogens within the cell. Lower levels of reactive oxygen species initiate systemic acquired resistance, triggering defence-molecule production in the rest of the plant.

In some plants, chloroplasts are known to move closer to the infection site and the nucleus during an infection.

Chloroplasts can serve as cellular sensors. After detecting stress in a cell, which might be due to a pathogen, chloroplasts begin producing molecules like salicylic acid, jasmonic acid, nitric oxide and reactive oxygen species which can serve as defence-signals. As cellular signals, reactive oxygen species are unstable molecules, so they probably don't leave the chloroplast, but instead pass on their signal to an unknown second messenger molecule. All these molecules initiate retrograde signalling—signals from the chloroplast that regulate gene expression in the nucleus.

In addition to defence signalling, chloroplasts, with the help of the peroxisomes, help synthesize an important defence molecule, jasmonate. Chloroplasts synthesize all the fatty acids in a plant cell—linoleic acid, a fatty acid, is a precursor to jasmonate.

Photosynthesis

One of the main functions of the chloroplast is its role in photosynthesis, the process by which light is transformed into chemical energy, to subsequently produce food in the form of sugars. Water (H_2O) and carbon dioxide (CO_2) are used in photosynthesis, and sugar and oxygen (O_2) is made, using light energy. Photosynthesis is divided into two stages—the light reactions, where water is split to produce oxygen, and the dark reactions, or Calvin cycle, which builds sugar molecules from carbon dioxide. The two phases are linked by the energy carriers adenosine triphosphate (ATP) and nicotinamide adenine dinucleotide phosphate ($NADP^+$).

Light Reactions

The light reactions of photosynthesis take place across the thylakoid membranes.

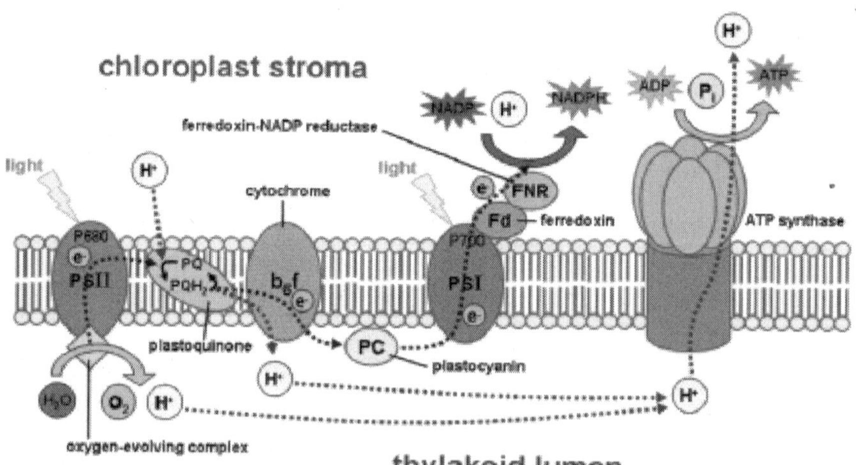

Figure: *The light reactions take place on the thylakoid membranes. They take light energy and store it in NADPH, a form of NADP⁺, and ATP to fuel the dark reactions.*

Energy Carriers

ATP is the phosphorylated version of adenosine diphosphate (ADP), which stores energy in a cell and powers most cellular activities. ATP is the energized form, while ADP is the (partially) depleted form.

$NADP^+$ is an electron carrier which ferries high energy electrons. In the light reactions, it gets reduced, meaning it picks up electrons, becoming NADPH.

Photophosphorylation

Like mitochondria, chloroplasts use the potential energy stored in an H^+, or hydrogen ion gradient to generate ATP energy. The two photosystems capture light energy to energize electrons taken from water, and release them down an electron transport chain. The molecules between the photosystems harness the electrons' energy to pump hydrogen ions into the thylakoid space, creating a concentration gradient, with more hydrogen ions (up to a thousand times as many) inside the thylakoid system than in the stroma. The hydrogen ions in the thylakoid space then diffuse back down their concentration gradient, flowing back out into the stroma through ATP synthase. ATP synthase uses the energy from the flowing hydrogen ions to phosphorylate adenosine diphosphate into adenosine triphosphate, or ATP. Because chloroplast ATP synthase projects out into the stroma, the ATP is synthesized there, in position to be used in the dark reactions.

$NADP^+$ Reduction

Electrons are often removed from the electron transport chains to charge $NADP^+$ with electrons, reducing it to NADPH. Like ATP synthase, ferredoxin-$NADP^+$ reductase, the enzyme that reduces $NADP^+$, releases the NADPH it makes into the stroma, right where it is needed for the dark reactions.

Because $NADP^+$ reduction removes electrons from the electron transport chains, they must be replaced—the job of photosystem II, which splits water molecules (H_2O) to obtain the electrons from its hydrogen atoms.

Cyclic Photosystem

While photosystem II photolyzes water to obtain and energize new electrons, photosystem I simply reenergizes depleted electrons at the end of an electron transport chain. Normally, the reenergized electrons are taken by $NADP^+$, though sometimes they can flow back down more H^+-pumping electron transport chains to transport more hydrogen ions into the thylakoid space to generate more ATP. This is termed cyclic photophosphorylation because the electrons are recycled. Cyclic photophosphorylation is common in C_4 plants, which need more ATP than NADPH.

Dark Reactions

The Calvin cycle, also known as the dark reactions, is a series of biochemical reactions that fixes CO_2 into G3P sugar molecules and uses the energy and electrons from the ATP and NADPH made in the light reactions. The Calvin cycle takes place in the stroma of the chloroplast.

While named *"the dark reactions"*, in most plants, they take place in the light, since the dark reactions are dependent on the products of the light reactions.

Carbon Fixation and G3P Synthesis

The Calvin cycle starts by using the enzyme Rubisco to fix CO_2 into five-carbon Ribulose bisphosphate (RuBP) molecules. The result is unstable six-carbon molecules that immediately break down into three-carbon molecules called 3-phosphoglyceric acid, or 3-PGA. The ATP and NADPH made in the light reactions is used to convert the 3-PGA into glyceraldehyde-3-phosphate, or G3P sugar molecules. Most of the G3P molecules are recycled back into RuBP using energy from more ATP, but one out of every six produced leaves the cycle—the end product of the dark reactions.

Sugars and Starches

Glyceraldehyde-3-phosphate can double up to form larger sugar molecules like glucose and fructose. These molecules are processed, and from them, the still larger sucrose, a disaccharide commonly known as table sugar, is made, though this process takes place outside of the chloroplast, in the cytoplasm.

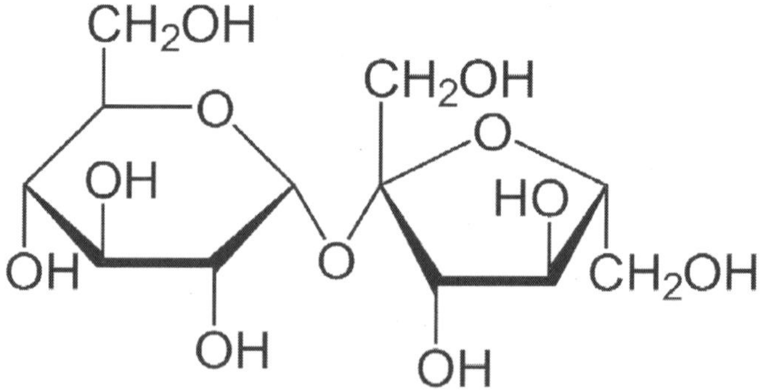

Figure: *Sucrose is made up of a glucose monomer (left), and a fructose monomer (right).*

Alternatively, glucose monomers in the chloroplast can be linked together to make starch, which accumulates into the starch grains found in the chloroplast. Under conditions such as high atmospheric CO_2 concentrations, these starch grains may grow very large, distorting the grana and thylakoids. The starch granules displace the thylakoids, but leave them intact. Waterlogged roots can also cause starch buildup in the chloroplasts, possibly due to less sucrose being exported out of the chloroplast (or more accurately, the plant cell).

This depletes a plant's free phosphate supply, which indirectly stimulates chloroplast starch synthesis. While linked to low photosynthesis rates, the starch grains themselves may not necessarily interfere significantly with the efficiency of photosynthesis, and might simply be a side effect of another photosynthesis-depressing factor.

Photorespiration

Photorespiration can occur when the oxygen concentration is too high. Rubisco cannot distinguish between oxygen and carbon dioxide very well, so it can accidentally add O_2 instead of CO_2 to RuBP. This process reduces the efficiency of photosynthesis—it consumes ATP and oxygen, releases CO_2, and produces no sugar. It can waste up to half the carbon fixed by the Calvin cycle. Several mechanisms have evolved in different lineages that raise the carbon dioxide concentration relative to oxygen within the chloroplast, increasing the efficiency of photosynthesis. These mechanisms are called carbon dioxide concentrating mechanisms, or CCMs. These include Crassulacean acid metabolism, C_4 carbon fixation, and pyrenoids. Chloroplasts in C_4 plants are notable as they exhibit a distinct chloroplast dimorphism.

pH

Because of the H^+ gradient across the thylakoid membrane, the interior of the thylakoid is acidic, with a pH around 4, while the stroma is slightly basic, with a pH of around 8. The optimal stroma pH for the Calvin cycle is 8.1, with the reaction nearly stopping when the pH falls below 7.3.

CO_2 in water can form carbonic acid, which can disturb the pH of isolated chloroplasts, interfering with photosynthesis, even though CO_2 is used in photosynthesis. However, chloroplasts in living plant cells are not affected by this as much.

Chloroplasts can pump K^+ and H^+ ions in and out of themselves using a poorly understood light-driven transport system.

In the presence of light, the pH of the thylakoid lumen can drop up to 1.5 pH units, while the pH of the stroma can rise by nearly one pH unit.

Amino Acid Synthesis

Chloroplasts alone make almost all of a plant cell's amino acids in their stroma except the sulfur-containing ones like cysteine and methionine. Cysteine is made in the chloroplast (the proplastid too) but it is also synthesized in the cytosol and mitochondria, probably because it has trouble crossing membranes to get to where it is needed. The chloroplast is known to make the precursors to methionine but it is unclear whether the organelle carries out the last leg of the pathway or if it happens in the cytosol.

Other Nitrogen Compounds

Chloroplasts make all of a cell's purines and pyrimidines—the nitrogenous bases found in DNA and RNA. They also convert nitrite (NO_2^+) into ammonia (NH_3) which supplies the plant with nitrogen to make its amino acids and nucleotides.

Other Chemical Products

Chloroplasts are the site of complex lipid metabolism.

Differentiation, Replication, and Inheritance

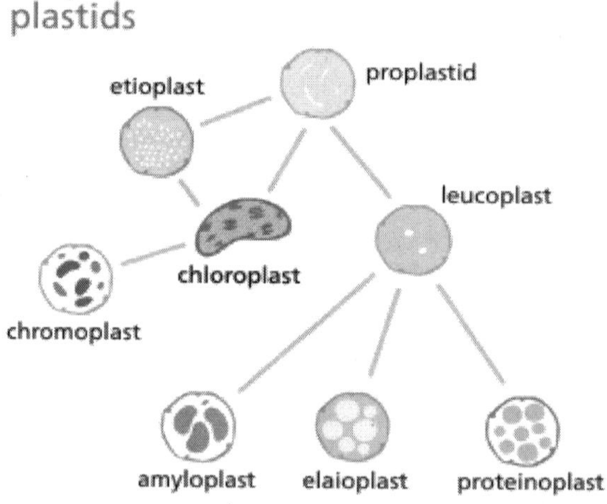

Figure: *Plastid types* (Interactive diagram) *Plants contain many different kinds of plastids in their cells.*

Chloroplasts are a special type of a plant cell organelle called a plastid, though the two terms are sometimes used interchangeably. There are many other types of plastids, which carry out various functions. All chloroplasts in a plant are descended from undifferentiated proplastids found in the zygote, or fertilized egg. Proplastids are commonly found in an adult plant's apical meristems.

Chloroplasts do not normally develop from proplastids in root tip meristems—instead, the formation of starch-storing amyloplasts is more common.

In shoots, proplastids from shoot apical meristems can gradually develop into chloroplasts in photosynthetic leaf tissues as the leaf matures, if exposed to the required light. This process involves invaginations of the inner plastid membrane, forming sheets of membrane that project into the internal stroma. These membrane sheets then fold to form thylakoids and grana.

If angiosperm shoots are not exposed to the required light for chloroplast formation, proplastids may develop into an etioplast stage before becoming chloroplasts. An etioplast is a plastid that lacks chlorophyll, and has inner membrane invaginations that form a lattice of tubes in their stroma, called a prolamellar body. While etioplasts lack chlorophyll, they have a yellow chlorophyll precursor stocked. Within a few minutes of light exposure, the prolamellar body begins to reorganise into stacks of thylakoids, and chlorophyll starts to be produced. This process, where the etioplast becomes a chloroplast, takes several hours. Gymnosperms do not require light to form chloroplasts.

Light, however, does not guarantee that a proplastid will develop into a chloroplast. Whether a proplastid develops into a chloroplast some other kind of plastid is mostly controlled by the nucleus and is largely influenced by the kind of cell it resides in.

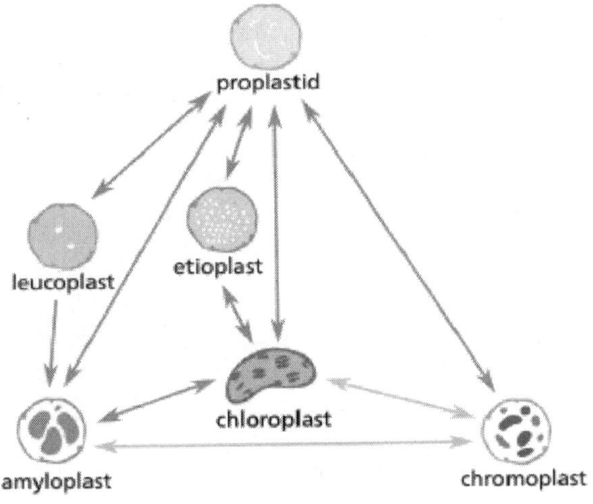

Figure: *Many plastid interconversions are possible.*

Plastid Interconversion

Plastid differentiation is not permanent, in fact many interconversions are possible. Chloroplasts may be converted to chromoplasts, which are pigment-filled plastids responsible for the bright colours you see in flowers and ripe fruit. Starch storing amyloplasts can also be converted to chromoplasts, and it is possible for proplastids to develop straight into chromoplasts. Chromoplasts and amyloplasts can also become chloroplasts, like what happens when you illuminate a carrot or a potato. If a plant is injured, or something else causes a plant cell to revert to a meristematic state, chloroplasts and other plastids can turn back into proplastids. Chloroplast, amyloplast, chromoplast, proplast, etc., are not absolute states—intermediate forms are common.

Chloroplast Division

Most chloroplasts in a photosynthetic cell do not develop directly from proplastids or etioplasts. In fact, a typical shoot meristematic plant cell contains only 7–20 proplastids. These proplastids differentiate into chloroplasts, which divide to create the 30–70 chloroplasts found in a mature photosynthetic plant cell. If the cell divides, chloroplast division provides the additional chloroplasts to partition between the two daughter cells.

In single-celled algae, chloroplast division is the only way new chloroplasts are formed. There is no proplastid differentiation—when an algal cell divides, its chloroplast divides along with it, and each daughter cell receives a mature chloroplast.

Almost all chloroplasts in a cell divide, rather than a small group of rapidly dividing chloroplasts. Chloroplasts have no definite S-phase—their DNA replication is not synchronized or limited to that of their host cells. Much of what we know about chloroplast division comes from studying organisms like *Arabidopsis* and the red alga *Cyanidioschyzon merolæ*.

The division process starts when the proteins FtsZ1 and FtsZ2 assemble into filaments, and with the help of a protein ARC6, form a structure called a Z-ring within the chloroplast's stroma. The Min system manages the placement of the Z-ring, ensuring that the chloroplast is cleaved more or less evenly. The protein MinD prevents FtsZ from linking up and forming filaments. Another protein ARC3 may also be involved, but it is not very well understood. These proteins are active at the poles of the chloroplast, preventing Z-ring formation there, but near the centre of the chloroplast, MinE inhibits them, allowing the Z-ring to form. Next, the two plastid-dividing rings, or PD rings form. The inner plastid-dividing ring is located in the inner side of the chloroplast's inner membrane, and is formed first. The outer plastid-dividing ring is found wrapped around the outer chloroplast membrane. It consists of filaments about 5 nanometres across, arranged in rows 6.4 nanometres apart, and shrinks to squeeze the chloroplast. This is when chloroplast constriction begins.

In a few species like *Cyanidioschyzon merolæ*, chloroplasts have a third plastid-dividing ring located in the chloroplast's intermembrane space.

Late into the constriction phase, dynamin proteins assemble around the outer plastid-dividing ring, helping provide force to squeeze the chloroplast. Meanwhile, the Z-ring and the inner plastid-dividing ring break down. During this stage, the many chloroplast DNA plasmids floating around in the stroma are partitioned and distributed to the two forming daughter chloroplasts.

Later, the dynamins migrate under the outer plastid dividing ring, into direct contact with the chloroplast's outer membrane, to cleave the chloroplast in two daughter chloroplasts.

A remnant of the outer plastid dividing ring remains floating between the two daughter chloroplasts, and a remnant of the dynamin ring remains attached to one of the daughter chloroplasts.

Of the five or six rings involved in chloroplast division, only the outer plastid-dividing ring is present for the entire constriction and division phase—while the Z-ring forms first, constriction does not begin until the outer plastid-dividing ring forms.

Regulation

In species of algae which contain a single chloroplast, regulation of chloroplast division is extremely important to ensure that each daughter cell receives a chloroplast—chloroplasts can't be made from scratch. In organisms like plants, whose cells contain multiple chloroplasts, coordination is looser and less important. It is likely that chloroplast and cell division are somewhat synchronized, though the mechanisms for it are mostly unknown.

Light has been shown to be a requirement for chloroplast division. Chloroplasts can grow and progress through some of the constriction stages under poor quality green light, but are slow to complete division—they require exposure to bright white light to complete division. Spinach leaves grown under green light have been observed to contain many large dumbbell-shaped chloroplasts. Exposure to white light can stimulate these chloroplasts to divide and reduce the population of dumbbell-shaped chloroplasts.

Chloroplast Inheritance

Like mitochondria, chloroplasts are usually inherited from a single parent. Biparental chloroplast inheritance—where plastid genes are inherited from both parent plants—occurs in very low levels in some flowering plants.

Many mechanisms prevent biparental chloroplast DNA inheritance including selective destruction of chloroplasts or their genes within the gamete or zygote, and chloroplasts from one parent being excluded from the embryo. Parental chloroplasts can be sorted so that only one type is present in each offspring.

Gymnosperms, such as pine trees, mostly pass on chloroplasts paternally, while flowering plants often inherit chloroplasts maternally. Flowering plants were once thought to only inherit chloroplasts maternally. However, there are now many documented cases of angiosperms inheriting chloroplasts paternally.

Angiosperms which pass on chloroplasts maternally have many ways to prevent paternal inheritance. Most of them produce sperm cells which do not contain any plastids. There are many other documented mechanisms that prevent paternal inheritance in these flowering plants, such as different rates of chloroplast replication within the embryo.

Among angiosperms, paternal chloroplast inheritance is observed more often in hybrids than in offspring from parents of the same species. This suggests that incompatible hybrid genes might interfere with the mechanisms that prevent paternal inheritance.

Transplastomic Plants

Recently, chloroplasts have caught attention by developers of genetically modified crops. Since in most flowering plants, chloroplasts are not inherited from the male parent, transgenes in these plastids cannot be disseminated by pollen. This makes plastid transformation a valuable tool for the creation and cultivation of genetically modified plants that are biologically contained, thus posing significantly lower environmental risks.

This biological containment strategy is therefore suitable for establishing the coexistence of conventional and organic agriculture. While the reliability of this mechanism has not yet been studied for all relevant crop species, recent results in tobacco plants are promising, showing a failed containment rate of transplastomic plants at 3 in 1,000,000.

Thylakoid

A thylakoid is a membrane-bound compartment inside chloroplasts and cyanobacteria. They are the site of the light-dependent reactions of photosynthesis. Thylakoids consist of a thylakoid membrane surrounding a thylakoid lumen. Chloroplast thylakoids frequently form stacks of disks referred to as grana (singular: granum). Grana are connected by intergranal or stroma thylakoids, which join granum stacks together as a single functional compartment.

Etymology

The word *thylakoid* come via Latin from Greek *thylakos* meaning "sac" or "pouch". Thus, *thylakoid* means "sac-like" or "pouch-like".

Thylakoid Structure

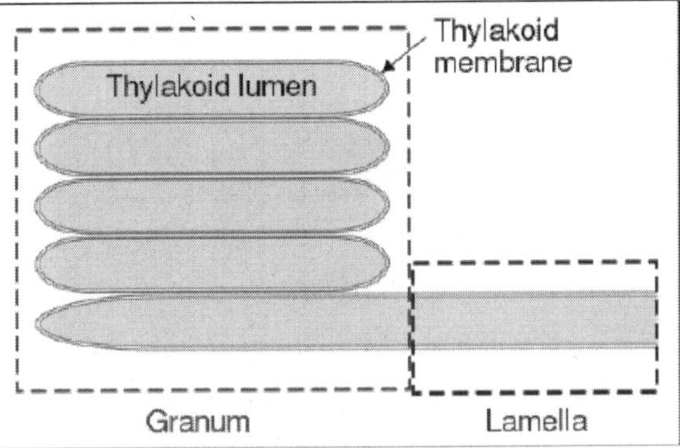

Figure: *Thylakoid structures*

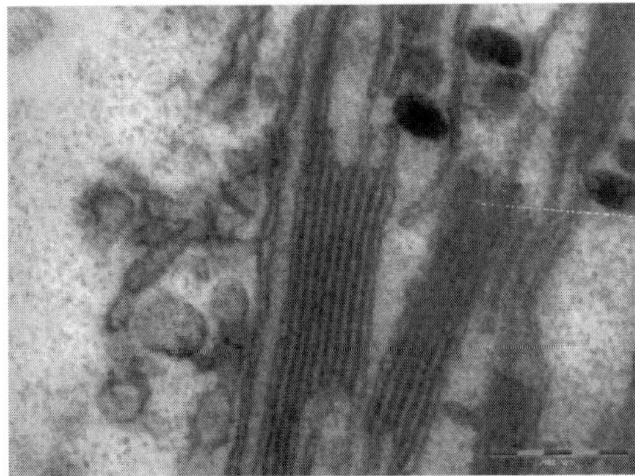

Figure: *TEM image of grana*

Thylakoids are membrane-bound structures embedded into the chloroplast stroma. A stack of thylakoids in a granum resembles a stack of coins.

Membrane

The thylakoid membrane is the site of the light-dependent reactions of photosynthesis with the photosynthetic pigments embedded directly in the membrane. It is an alternating pattern of dark and light bands measuring each 1 nanometre. The thylakoid lipid bilayer shares characteristic features with prokaryotic membranes and the inner chloroplast membrane.

For example, acidic lipids can be found in thylakoid membranes, cyanobacteria and other photosynthetic bacteria and are involved in the functional integrity of the photosystems. The thylakoid membranes of higher plants are composed primarily of phospholipids and galactolipids that are asymmetrically arranged along and across the membranes. Thylakoid membranes are richer in galactolipids rather than phospholipids; also they predominantly consist of hexagonal phase II forming monogalacotosyl diglyceride lipid. Despite this unique composition, plant thylakoid membranes have been shown to assume largely lipid-bilayer dynamic organisation. The lipids for the thylakoid membranes, richest in high-fluidity linolenic acid are synthesized in a complex pathway involving exchange of lipid precursors between the endoplasmic reticulum and inner membrane of the plastid envelope and transported from the inner membrane to the thylakoids via vesicles.

Lumen

The thylakoid lumen is a continuous aqueous phase enclosed by the thylakoid membrane. It plays a vital role for photophosphorylation during photosynthesis. During the light-dependent reaction, protons are pumped across the thylakoid membrane into the lumen making it acidic down to pH 4.

Granum and Stroma Lamellae

In higher plants thylakoids are organised into a granum-stroma membrane assembly. A granum (plural grana) is a stack of thylakoid discs. Chloroplasts can have from 10 to 100 grana. Grana are connected by stroma thylakoids, also called intergrana thylakoids or lamellae. Grana thylakoids and stroma thylakoids can be distinguished by their different protein composition. Grana contribute to chloroplasts' large surface area to volume ratio. Different interpretations of electron tomography imaging of thylakoid membranes has resulted in two models for grana structure. Both posit that lamellae intersect grana stacks in parallel sheets, though whether these sheets intersect in planes perpendicular to the grana stack axis, or are arranged in a right-handed helix is debated.

Thylakoid Formation

Chloroplasts develop from proplastids when seedlings emerge from the ground. Thylakoid formation requires light. In the plant embryo and in the absence of light, proplastids develop into etioplasts that contain semicrystalline membrane structures called prolamellar bodies. When exposed to light, these prolamellar bodies develop into thylakoids. This does not happen in seedlings grown in the dark, which undergo etiolation. An underexposure to light can cause the thylakoids to fail. This causes the chloroplasts to fail resulting in the death of the plant.

Thylakoid formation requires the action of *vesicle-inducing protein in plastids 1* (VIPP1). Plants cannot survive without this protein, and reduced VIPP1 levels lead to slower growth and paler plants with reduced ability to photosynthesize. VIPP1 appears to be required for basic thylakoid membrane formation, but not for the assembly of protein complexes of the thylakoid membrane. It is conserved in all organisms containing thylakoids, including cyanobacteria, green algae, such as Chlamydomonas, and higher plants, such as *Arabidopsis thaliana*.

Thylakoid Isolation and Fractionation

Thylakoids can be purified from plant cells using a combination of differential and gradient centrifugation. Disruption of isolated thylakoids, for example by mechanical shearing, releases the lumenal fraction. Peripheral and integral membrane fractions can be extracted from the remaining membrane fraction. Treatment with sodium carbonate (Na_2CO_3) detaches peripheral membrane proteins, whereas treatment with detergents and organic solvents solubilizes integral membrane proteins.

Thylakoid Proteins

Thylakoids contain many integral and peripheral membrane proteins, as well as lumenal proteins. Recent proteomics studies of thylakoid fractions have provided further details on the protein composition of the thylakoids. These data have been summarized in several plastid protein databases that are available online.

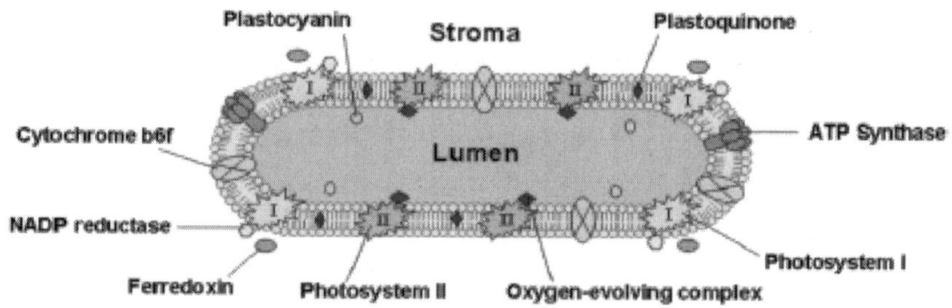

Figure: *Thylakoid disc with embedded and associated proteins*

According to these studies, the thylakoid proteome consists of at least 335 different proteins. Out of these, 89 are in the lumen, 116 are integral membrane proteins, 62 are peripheral proteins on the stroma side, and 68 peripheral proteins on the lumenal side. Additional low-abundance lumenal proteins can be predicted through computational methods. Of the thylakoid proteins with known functions, 42% are involved in photosynthesis. The next largest functional groups include proteins involved in protein targeting, processing and folding with 11%, oxidative stress response (9%) and translation (8%).

Integral Membrane Proteins

Thylakoid membranes contain integral membrane proteins which play an important role in light harvesting and the light-dependent reactions of photosynthesis. There are four major protein complexes in the thylakoid membrane:

- Photosystems I and II
- Cytochrome b6f complex
- ATP synthase

Photosystem II is located mostly in the grana thylakoids, whereas photosystem I and ATP synthase are mostly located in the stroma thylakoids and the outer layers of grana. The cytochrome b6f complex is distributed evenly throughout thylakoid membranes. Due to the separate location of the two photosystems in the thylakoid membrane system, mobile electron carriers are required to shuttle electrons between them. These carriers are plastoquinone and plastocyanin. Plastoquinone shuttles electrons from photosystem II to the cytochrome b6f complex, whereas plastocyanin carries electrons from the cytochrome b6f complex to photosystem I.

Together, these proteins make use of light energy to drive electron transport chains that generate a chemiosmotic potential across the thylakoid membrane and NADPH, a product of the terminal redox reaction. The ATP synthase uses the chemiosmotic potential to make ATP during photophosphorylation.

Photosystems

These photosystems are light-driven redox centres, each consisting of an antenna complex that uses chlorophylls and accessory photosynthetic pigments such as carotenoids and phycobiliproteins to harvest light at a variety of wavelengths. Each antenna complex has between 250 and 400 pigment molecules and the energy they absorb is shuttled by resonance energy transfer to a specialised chlorophyll *a* at the reaction centre of each photosystem. When either of the two chlorophyll *a* molecules at the reaction centre absorbs energy, an electron is excited and transferred to an electron-acceptor molecule. Photosystem I contains a pair of chlorophyll *a* molecules, designated P700, at its reaction centre that maximally absorbs 700 nm light. Photosystem II contains P680 chlorophyll that absorbs 680 nm light best (note that these wavelengths correspond to deep red). The P is short for pigment and the number is the specific absorption peak in nanometres for the chlorophyll molecules in each reaction centre.

Cytochrome b6f Complex

The cytochrome b6f complex is part of the thylakoid electron transport chain and couples electron transfer to the pumping of protons into the thylakoid lumen. Energetically, it is situated between the two photosystems and transfers electrons from photosystem II-plastoquinone to plastocyanin-photosystem I.

ATP Synthase

The thylakoid ATP synthase is a CF1FO-ATP synthase similar to the mitochondrial ATPase. It is integrated into the thylakoid membrane with the CF1-part sticking into stroma. Thus, ATP synthesis occurs on the stromal side of the thylakoids where the ATP is needed for the light-independent reactions of photosynthesis.

Thylakoid Lumen Proteins

The electron transport protein plastocyanin is present in the lumen and shuttles electrons from the cytochrome b6f protein complex to photosystem I. While plastoquinones are lipid-soluble and therefore move within the thylakoid membrane, plastocyanin moves through the thylakoid lumen.

The lumen of the thylakoids is also the site of water oxidation by the oxygen evolving complex associated with the lumenal side of photosystem II.

Lumenal proteins can be predicted computationally based on their targeting signals. In Arabidopsis, out of the predicted lumenal proteins possessing the Tat signal, the largest groups with known functions are 19% involved in protein processing (proteolysis and folding), 18% in photosynthesis, 11% in metabolism, and 7% redox carriers and defence.

Thylakoid Protein Expression

Chloroplasts have their own genome, which encodes a number of thylakoid proteins. However, during the course of plastid evolution from their cyanobacterial endosymbiotic ancestors, extensive gene transfer from the chloroplast genome to the cell nucleus took place. This results in the four major thylakoid protein complexes being encoded in part by the chloroplast genome and in part by the nuclear genome. Plants have developed several mechanisms to co-regulate the expression of the different subunits encoded in the two different organelles to assure the proper stoichiometry and assembly of these protein complexes. For example, transcription of nuclear genes encoding parts of the photosynthetic apparatus is regulated by light. Biogenesis, stability and turnover of thylakoid protein complexes are regulated by phosphorylation via redox-sensitive kinases in the thylakoid membranes.

The translation rate of chloroplast-encoded proteins is controlled by the presence or absence of assembly partners (control by epistasy of synthesis). This mechanism involves negative feedback through binding of excess protein to the 5' untranslated region of the chloroplast mRNA. Chloroplasts also need to balance the ratios of photosystem I and II for the electron transfer chain. The redox state of the electron carrier plastoquinone in the thylakoid membrane directly affects the transcription of chloroplast genes encoding proteins of the reaction centres of the photosystems, thus counteracting imbalances in the electron transfer chain.

Protein Targeting to the Thylakoids

Thylakoid proteins are targeted to their destination via signal peptides and prokaryotic-type secretory pathways inside the chloroplast. Most thylakoid proteins encoded by a plant's nuclear genome need two targeting signals for proper localization: An N-terminal chloroplast targeting peptide, followed by a thylakoid targeting peptide (shown in blue). Proteins are imported through the translocon of outer and inner membrane (Toc and Tic) complexes. After entering the chloroplast, the first targeting peptide is cleaved off by a protease processing imported proteins. This unmasks the second targeting signal and the protein is exported from the stroma into the thylakoid in a second targeting step. This second step requires the action of protein translocation components of the thylakoids and is energy-dependent. Proteins are inserted into the membrane via the SRP-dependent pathway (1), the Tat-dependent pathway (2), or spontaneously via their transmembrane domains. Lumenal proteins are exported across the thylakoid membrane into the lumen by either the Tat-dependent pathway (2) or the Sec-dependent pathway (3) and released by cleavage from the thylakoid targeting signal. The different pathways utilise different signals and energy sources. The Sec (secretory) pathway requires ATP as energy source and consists of SecA, which binds to the imported protein and a Sec membrane complex to shuttle the protein across. Proteins with a twin arginine motif in their thylakoid signal peptide are shuttled through the Tat (twin arginine translocation) pathway, which requires a membrane-bound Tat complex and the pH gradient as an energy source. Some other proteins are inserted into the membrane via the SRP (signal recognition particle) pathway. The chloroplast SRP can interact with its target proteins either post-translationally or co-translationally, thus transporting imported proteins as well as those that are translated inside the chloroplast. The SRP pathway requires GTP and the pH gradient as energy sources. Some transmembrane proteins may also spontaneously insert into the membrane from the stromal side without energy requirement.

Thylakoid Function

The thylakoids are the site of the light-dependent reactions of photosynthesis. These include light-driven water oxidation and oxygen evolution, the pumping of protons across the thylakoid membranes coupled with the electron transport chain of the photosystems and cytochrome b6f complex, and ATP synthesis by the ATP synthase utilising the generated proton gradient.

Water Photolysis

The first step in photosynthesis is the light-driven reduction (splitting) of water to provide the electrons for the photosynthetic electron transport chains as well as protons for the establishment of a proton gradient. The water-splitting reaction occurs on the lumenal side of the thylakoid membrane and is driven by the light energy captured by the photosystems. It is interesting to note that this reduction of water conveniently produces the waste product O_2 that is vital for cellular respiration. The molecular oxygen formed by the reaction is released into the atmosphere.

Electron Transport Chains

Two different variations of electron transport are used during photosynthesis:

- Noncyclic electron transport or Non-cyclic photophosphorylation produces NADPH + H^+ and ATP.
- Cyclic electron transport or Cyclic photophosphorylation produces only ATP.

The noncyclic variety involves the participation of both photosystems, while the cyclic electron flow is dependent on only photosystem I.

- Photosystem I uses light energy to reduce $NADP^+$ to NADPH + H^+, and is active in both noncyclic and cyclic electron transport. In cyclic mode, the energized electron is passed down a chain that ultimately returns it (in its base state) to the chlorophyll that energized it.
- Photosystem II uses light energy to oxidize water molecules, producing electrons (e^-), protons (H^+), and molecular oxygen (O_2), and is only active in noncyclic transport. Electrons in this system are not conserved, but are rather continually entering from oxidized $2H_2O$ (O_2 + 4 H^+ + 4 e^-) and exiting with $NADP^+$ when it is finally reduced to NADPH.

Chemiosmosis

A major function of the thylakoid membrane and its integral photosystems is the establishment of chemiosmotic potential. The carriers in the electron transport chain use some of the electron's energy to actively transport protons from the stroma to the lumen. During photosynthesis, the lumen becomes acidic, as low as pH 4, compared to pH 8 in the stroma. This represents a 10,000 fold concentration gradient for protons across the thylakoid membrane.

Source of Proton Gradient

The protons in the lumen come from three primary sources.

- Photolysis by photosystem II oxidises water to oxygen, protons and electrons in the lumen.
- The transfer of electrons from photosystem II to plastoquinone during non-cyclic electron transport consumes two protons from the stroma. These are released in the lumen when the

reduced plastoquinol is oxidized by the cytochrome b6f protein complex on the lumen side of the thylakoid membrane. From the plastoquinone pool, electrons pass through the cytochrome b6f complex. This integral membrane assembly resembles cytochrome bc1.

- The reduction of plastoquinone by ferredoxin during cyclic electron transport also transfers two protons from the stroma to the lumen.

The proton gradient is also caused by the consumptions of protons in the stroma to make NADPH from NADP+ at the NADP reductase.

ATP Generation

The molecular mechanism of ATP generation in chloroplasts is similar to that in mitochondria and takes the required energy from the proton motive force (PMF). However, chloroplasts rely more on the chemical potential of the PMF to generate the potential energy required for ATP synthesis. The PMF is the sum of a proton chemical potential (given by the proton concentration gradient) and a transmembrane electrical potential (given by charge separation across the membrane). Compared to the inner membranes of mitochondria, which have a significantly higher membrane potential due to charge separation, thylakoid membranes lack a charge gradient. To compensate for this, the 10,000 fold proton concentration gradient across the thylakoid membrane is much higher compared to a 10 fold gradient across the inner membrane of mitochondria. The resulting chemiosmotic potential between the lumen and stroma is high enough to drive ATP synthesis using the ATP synthase. As the protons travel back down the gradient through channels in ATP synthase, ADP + P_i are combined into ATP. In this manner, the light-dependent reactions are coupled to the synthesis of ATP via the proton gradient.

Thylakoid Membranes in Cyanobacteria

Cyanobacteria are photosynthetic prokaryotes with highly differentiated membrane systems. Cyanobacteria have an internal system of thylakoid membranes where the fully functional electron transfer chains of photosynthesis and respiration reside. The presence of different membrane systems lends these cells a unique complexity among bacteria. Cyanobacteria must be able to reorganise the membranes, synthesize new membrane lipids, and properly target proteins to the correct membrane system. The outer membrane, plasma membrane, and thylakoid membranes each have specialised roles in the cyanobacterial cell. Understanding the organisation, functionality, protein composition and dynamics of the membrane systems remains a great challenge in cyanobacterial cell biology.

The thylakoid membranes of the cyanobacteria are not differentiated into granum and stroma regions as it happens in plants. They form stacks of parallel sheets close to the cytoplasmic membrane with a low packing density. The relatively large distance between the thylakoids provides space for the external light harvesting antennae, the phycobilisomes. This macrostructure, as in the case of higher plants, shows some flexibility during changes in the physicochemical environment.

Light-Dependent Reactions

The light-dependent reactions, or photoreduction, is the first stage of photosynthesis, is a process by which plants capture and store energy from sunlight. In this process, light energy is converted into chemical energy, in the form of the energy-carrying molecules ATP and NADPH. In the *light-independent reactions*, the formed NADPH and ATP drive the reduction of CO_2 to more useful organic compounds, such as glucose. However, although light-independent reactions are, by convention, also called *dark reactions*, they are not independent of the need of light, for they are driven by ATP and NADPH, products of light. They are often called the Calvin Cycle or C3 Cycle.

The light-dependent reactions take place on the thylakoid membranes. The inside of the thylakoid membrane is called the lumen, and outside the thylakoid membrane is the stroma, where the light-independent reactions take place. The thylakoid membrane contains some integral membrane protein complexes that catalyze the light reactions. There are four major protein complexes in the thylakoid membrane: Photosystem II (PSII), Cytochrome b6f complex, Photosystem I (PSI), and ATP synthase. These four complexes work together to ultimately create the products ATP and NADPH.

The two photosystems absorb light energy through pigments - primarily the chlorophylls, which are responsible for the green colour of leaves. The light-dependent reactions begin in photosystem II. When a chlorophyll *a* molecule within the reaction centre of PSII absorbs a photon, an electron in this molecule attains a higher energy level. Because this state of an electron is very unstable, the electron is transferred from one to another molecule creating a chain of redox reactions, called an electron transport chain (ETC). The electron flow goes from PSII to cytochrome b6f to PSI. In PSI, the electron gets the energy from another photon. The final electron acceptor is NADP. In *oxygenic photosynthesis*, the first electron donor is water, creating oxygen as a waste product. In *anoxygenic photosynthesis* various electron donors are used.

Cytochrome b6f and ATP synthase work together to create ATP. This process is called photophosphorylation, which occurs in two different ways. In *non-cyclic photophosphorylation*, cytochrome b6f uses the energy of electrons from PSII to pump protons from the stroma to the lumen. The proton gradient across the thylakoid membrane creates a proton-motive force, used by ATP synthase to form ATP. In *cyclic photophosphorylation*, cytochrome b6f uses the energy of electrons from not only PSII but also PSI to create more ATP and to stop the production of NADPH. Cyclic phosphorylation is important to create ATP and maintain NADPH in the right proportion for the light-independent reactions.

The net-reaction of all light-dependent reactions in oxygenic photosynthesis is:

$$2H_2O + 2NADP + 3ADP + 3P_i \rightarrow O_2 + 2NADPH + 3ATP$$

The two photosystems are protein complexes that absorb photons and are able to use this energy to create an electron transport chain. Photosystem I and II are very similar in structure and function. They use special proteins, called light-harvesting complexes, to absorb the photons with very high effectiveness. If a special pigment molecule in a photosynthetic reaction centre absorbs a photon, an electron in this pigment attains the excited state and then is transferred to another molecule in the reaction centre. This reaction, called *photoinduced charge separation*, is the start of the electron flow and is unique because it transforms light energy into chemical forms.

The Reaction Centre

The reaction centre is in the thylakoid membrane. It transfers light energy to a dimer of chlorophyll pigment molecules near the periplasmic (or thylakoid lumen) side of the membrane. This dimer is called a special pair because of its fundamental role in photosynthesis. This special pair is slightly different in PSI and PSII reaction centre. In PSII, it absorbs photons with a wavelength of 680 nm, and it is therefore called P680. In PSI, it absorbs photons at 700 nm, and it is called P700. In bacteria, the special pair is called P760, P840, P870, or P960.

If an electron of the special pair in the reaction centre becomes excited, it cannot transfer this energy to another pigment using resonance energy transfer. In normal circumstances, the electron should return to the ground state, but, because the reaction centre is arranged so that a suitable electron acceptor is nearby, the excited electron can move from the initial molecule to the acceptor.

This process results in the formation of a positive charge on the special pair (due to the loss of an electron) and a negative charge on the acceptor and is, hence, referred to as photoinduced charge separation. In other words, electrons in pigment molecules can exist at specific energy levels. Under normal circumstances, they exist at the lowest possible energy level they can. However, if there is enough energy to move them into the next energy level, they can absorb that energy and occupy that higher energy level.

The light they absorb contains the necessary amount of energy needed to push them into the next level. Any light that does not have enough or has too much energy cannot be absorbed and is reflected. The electron in the higher energy level, however, does not want to be there; the electron is unstable and must return to its normal lower energy level. To do this, it must release the energy that has put it into the higher energy state to begin with. This can happen various ways.

The extra energy can be converted into molecular motion and lost as heat. Some of the extra energy can be lost as heat energy, while the rest is lost as light. This re-emission of light energy is called fluorescence. The energy, but not the e- itself, can be passed onto another molecule. This is called resonance. The energy and the e- can be transferred to another molecule. Plant pigments usually utilise the last two of these reactions to convert the sun's energy into their own.

This initial charge separation occurs in less than 10 picoseconds ($10^{"11}$ seconds). In their high-energy states, the special pigment and the acceptor could undergo charge recombination; that is, the electron on the acceptor could move back to neutralize the positive charge on the special pair. Its return to the special pair would waste a valuable high-energy electron and simply convert the absorbed light energy into heat. In the case of PSII, this backflow of electrons can produce reactive oxygen species leading to photoinhibition. Three factors in the structure of the reaction centre work together to suppress charge recombination nearly completely.

- Another electron acceptor is less than 10 Å away from the first acceptor, and so the electron is rapidly transferred farther away from the special pair.

- An electron donor is less than 10 Å away from the special pair, and so the positive charge is neutralized by the transfer of another electron
- The electron transfer back from the electron acceptor to the positively charged special pair is especially slow. The rate of an of electron transfer reaction increases with its thermodynamic favourability up to a point and then decreases. The back transfer is so favourable that it takes place in the inverted region where electron-transfer rates become slower.

Thus, electron transfer proceeds efficiently from the first electron acceptor to the next, creating an electron transport chain that ends if it has reached NADPH.

Photosynthetic Electron Transport Chains in Chloroplasts

The photosynthesis process in chloroplasts begins when an electron of P680 of PSII attains an higher-energy level. This energy is used to reduce a chain of electron acceptors that have subsequently lowered redox-potentials. This chain of electron acceptors is known as an electron transport chain. When this chain reaches PS I, an electron is again excited, creating a high redox-potential. The electron transport chain of photosynthesis is often put in a diagram called the z-scheme, because the redox diagram from P680 to P700 resembles the letter z.

The final product of PSII is plastoquinol, a mobile electron carrier in the membrane. Plastoquinol transfers the electron from PSII to the proton pump, cytochrome b6f. The ultimate electron donor of PSII is water. Cytochrome b6f proceeds the electron chain to PSI through plastocyanin molecules. PSI is able to continue the electron transfer in two different ways. It can transfer the electrons either to plastoquinol again, creating a cyclic electron flow, or to an enzyme called FNR (Ferredoxin—NADP(+) reductase), creating a non-cyclic electron flow. PSI releases FNR into the stroma, where it reduces NADP+.

to NADPH.

Activities of the electron transport chain, especially from cytochrome b6f, lead to pumping of protons from the stroma to the lumen. The resulting transmembrane proton gradient is used to make ATP via ATP synthase.

The overall process of the photosynthetic electron transport chain in chloroplasts is:

H_2O → PS II → plastoquinone → $cyt^b{}_6$ → plastocyanin → PS I → NADPH

Photosystem II

PS II is an extremely complex, highly organised transmembrane structure that contains a *water-splitting complex,* chlorophylls and carotenoid pigments, a *reaction centre* (P680), pheophytin (a pigment similar to chlorophyll), and two quinones. It uses the energy of sunlight to transfer electrons from water to a mobile electron carrier in the membrane called *plastoquinone*:

$H_2O \rightarrow P680 \rightarrow P680^* \rightarrow$ plastoquinone

Plastoquinone, in turn, transfers electrons to $cyt^b{}_6$, which feeds them into PS I.

The Water-Splitting Complex

The step $H_2O \rightarrow P680$ is performed by a poorly understood structure embedded within PS II called the *water-splitting complex* or the *oxygen-evolving complex.* It catalyzes a reaction that splits water into electrons, protons and oxygen:

$2H_2O \rightarrow 4H^+ + 4e^- + O_2$

The electrons are transferred to special chlorophyll molecules (embedded in PS II) that are promoted to a higher-energy state by the energy of photons.

The Reaction Centre

The excitation $P680 \rightarrow P680^*$ of the reaction centre pigment P680 occurs here. These special chlorophyll molecules embedded in PS II absorb the energy of photons, with maximal absorption at 680 nm. Electrons within these molecules are promoted to a higher-energy state. This is one of two core processes in photosynthesis, and it occurs with astonishing efficiency (greater than 90%) because, in addition to direct excitation by light at 680 nm, the energy of light first harvested by *antenna proteins* at other wavelengths in the light-harvesting system is also transferred to these special chlorophyll molecules.

This is followed by the step $P680^* \rightarrow$ pheophytin, and then on to plastoquinone, which occurs within the reaction centre of PS II. High-energy electrons are transferred to plastoquinone before it subsequently picks up two protons to become plastoquinol. Plastoquinol is then released into the membrane as a mobile electron carrier.

This is the second core process in photosynthesis. The initial stages occur within *picoseconds,* with an efficiency of 100%. The seemingly impossible efficiency is due to the precise positioning of molecules within the reaction centre. This is a solid-state process, not a chemical reaction. It occurs within an essentially crystalline environment created by the macromolecular structure of PS II. The usual rules of chemistry (which involve random collisions and random energy distributions) do not apply in solid-state environments.

Link of Water-Splitting Complex and Chlorophyll Excitation

When the chlorophyll passes the electron to pheophytin, it obtains an electron from P_{680}^{*}. In turn, P_{680}^{*} can oxidize the Z (or Y_Z) molecule. Once oxidized, the Z molecule can derive electrons from the water-splitting complex.

Summary

PS II is a transmembrane structure found in all chloroplasts. It splits water into electrons, protons and molecular oxygen. The electrons are transferred to plastoquinone, which carries them to a proton pump. Molecular oxygen is released into the atmosphere.

The emergence of such an incredibly complex structure, a macromolecule that converts the energy of sunlight into potentially useful work with efficiencies that are impossible in ordinary experience, seems almost magical at first glance.

Thus, it is of considerable interest that, in essence, the same structure is found in *purple bacteria*.

Cytochrome $^{b}_{6}$

PS II and PS I are connected by a transmembrane proton pump, cytochrome $^{b}_{6}$ complex (plastoquinol—plastocyanin reductase; EC 1.10.99.1). Electrons from PS II are carried by plastoquinol to cyt$^{b}_{6}$, where they are removed in a stepwise fashion (reforming plastoquinone) and transferred to a water-soluble electron carrier called *plastocyanin*. This redox process is coupled to the pumping of four protons across the membrane.

The resulting proton gradient (together with the proton gradient produced by the water-splitting complex in PS II) is used to make ATP via ATP synthase.

The similarity in structure and function between cytochrome $^{b}_{6}$ (in chloroplasts) and cytochrome $^{bc}_{1}$ (*Complex III* in mitochondria) is striking. Both are transmembrane structures that remove electrons from a mobile, lipid-soluble electron carrier (plastoquinone in chloroplasts; ubiquinone in mitochondria) and transfer them to a mobile, water-soluble electron carrier (plastocyanin in chloroplasts; cytochrome *c* in mitochondria). Both are proton pumps that produce a transmembrane proton gradient.

Photosystem I

PS I accepts electrons from plastocyanin and transfers them either to NADPH (*noncyclic electron transport*) or back to cytochrome$^b{}_6$ (*cyclic electron transport*):

$$\begin{array}{ccccccccc} \text{plastocyanin} & \rightarrow & \text{P700} & \rightarrow & \text{P700}^* & \rightarrow & \text{FNR} & \rightarrow & \text{NADPH} \\ \uparrow & & & & \downarrow & & & & \\ b_6 & & \leftarrow & & \text{plastoquinone} & & & & \end{array}$$

PS I, like PS II, is a complex, highly organised transmembrane structure that contains antenna chlorophylls, a reaction centre (P700), phylloquinine, and a number of iron-sulfur proteins that serve as intermediate redox carriers.

The light-harvesting system of PS I uses multiple copies of the same transmembrane proteins used by PS II. The energy of absorbed light (in the form of delocalized, high-energy electrons) is funneled into the reaction centre, where it excites special chlorophyll molecules (P700, maximum light absorption at 700 nm) to a higher energy level. The process occurs with astonishingly high efficiency.

Electrons are removed from excited chlorophyll molecules and transferred through a series of intermediate carriers to *ferredoxin*, a water-soluble electron carrier. As in PS II, this is a solid-state process that operates with 100% efficiency.

There are two different pathways of electron transport in PS I. In *noncyclic electron transport*, ferredoxin carries the electron to the enzyme ferredoxin NADP+ oxidoreductase (FNR) that reduces NADP+ to NADPH. In *cyclic electron transport*, electrons from ferredoxin are transferred (via plastoquinone) to a proton pump, cytochrome $^b{}_6$. They are then returned (via plastocyanin) to P700.

NADPH and ATP are used to synthesize organic molecules from CO_2. The ratio of NADPH to ATP production can be adjusted by adjusting the balance between cyclic and noncyclic electron transport.

It is noteworthy that PS I closely resembles photosynthetic structures found in *green sulfur bacteria*, just as PS II resembles structures found in *purple bacteria*.

Photosynthetic Electron Transport Chains in Bacteria

PS II, PS I, and cytochrome$^b{}_6$ are found in chloroplasts. All plants and all photosynthetic algae contain chloroplasts, which produce NADPH and ATP by the mechanisms described above. In essence, the same transmembrane structures are also found in *cyanobacteria*.

Unlike plants and algae, cyanobacteria are prokaryotes. They do not contain chloroplasts. Rather, they bear a striking resemblance to chloroplasts themselves. This suggests that organisms resembling cyanobacteria were the evolutionary precursors of chloroplasts. One imagines primitive eukaryotic cells taking up cyanobacteria as intracellular symbionts.

Cyanobacteria

Cyanobacteria contain structures similar to PS II and PS I in chloroplasts.

Their light-harvesting system is different from that found in plants (they use *phycobilins*, rather than chlorophylls, as antenna pigments), but their electron transport chain

H_2O → PS II → plastoquinone → $^{b}_{6}$ → cytochrome $^{c}_{6}$ → PS I → ferredoxin → NADPH
↑ ↓
$^{b}_{6}$ ← plastoquinone

is, in essence, the same as the electron transport chain in chloroplasts. The mobile water-soluble electron carrier is cytochrome *c*

6 in cyanobacteria, plastocyanin in plants.

Cyanobacteria can also synthesize ATP by oxidative phosphorylation, in the manner of other bacteria. The electron transport chain is

NADH dehydrogenase → plastoquinone → $^{b}_{6}$ → cytochrome $^{c}_{6}$ → cytochrome $^{aa}_{3}$ → O_2

where the mobile electron carriers are plastoquinone and cytochrome $^{c}_{6}$, while the proton pumps are NADH dehydrogenase, $^{b}_{6}$ and cytochrome $^{aa}_{3}$.

Cyanobacteria are the only bacteria that produce oxygen during photosynthesis. Earth's primordial atmosphere was anoxic. Organisms like cyanobacteria produced our present-day oxygen-containing atmosphere.

The other two major groups of photosynthetic bacteria, purple bacteria and green sulfur bacteria, contain only a single photosystem and do not produce oxygen.

Purple Bacteria

Purple bacteria contain a single photosystem that is structurally related to PS II in cyanobacteria and chloroplasts:

P870 → P870* → ubiquinone → $^{bc}_{1}$ → cytochrome $^{c}_{2}$ → *P870*

This is a *cyclic* process in which electrons are removed from an excited chlorophyll molecule (*bacteriochlorophyll*; P870), passed through an electron transport chain to a proton pump (cytochrome *bc*

1 complex, similar but not identical to cytochrome *bc*

1 in chloroplasts), and then returned to the cholorophyll molecule. The result is a proton gradient, which is used to make ATP via ATP synthase. As in cyanobacteria and chloroplasts, this is a solid-state process that depends on the precise orientation of various functional groups within a complex transmembrane macromolecular structure.

To make NADPH, purple bacteria use an external electron donor (hydrogen, hydrogen sulfide, sulfur, sulfite, or organic molecules such as succinate and lactate) to feed electrons into a reverse electron transport chain.

Green Sulfur Bacteria

Green sulfur bacteria contain a photosystem that is analogous to PS I in chloroplasts:

$$\begin{array}{ccccccc} \text{P840} & \rightarrow & \text{P840}^* & \rightarrow & \text{ferredoxin} & \rightarrow & \text{NADH} \\ \uparrow & & & & & & \downarrow \\ \text{cyt } c_{553} & \leftarrow & bc_1 & \leftarrow & \text{menaquinone} & & \end{array}$$

There are two pathways of electron transfer. In *cyclic electron transfer*, electrons are removed from an excited chlorophyll molecule, passed through an electron transport chain to a proton pump, and then returned to the chlorophyll.

The mobile electron carriers are, as usual, a lipid-soluble quinone and a water-soluble cytochrome. The resulting proton gradient is used to make ATP.

In *noncyclic electron transfer*, electrons are removed from an excited chlorophyll molecule and used to reduce NAD^+ to NADH. The electrons removed from P840 must be replaced. This is accomplished by removing electrons from H_2S, which is oxidized to sulfur (hence the name "green *sulfur* bacteria").

Purple bacteria and green sulfur bacteria occupy relatively minor ecological niches in the present day biosphere. They are of interest because of their importance in precambrian ecologies, and because their methods of photosynthesis were the likely evolutionary precursors of those in modern plants.

History

The first ideas about light's being used in photosynthesis were proposed by Colin Flannery in 1779 who recognised it was sunlight falling on plants that was required, although Joseph Priestley had noted the production of oxygen without the association with light in 1772. Cornelius Van Niel proposed in 1931 that photosynthesis is a case of general mechanism where a photon of light is used to photo decompose a hydrogen donor and the hydrogen being used to reduce CO_2. Then in 1939 Robin Hill showed that isolated chloroplasts would make oxygen, but not fix CO_2 showing the light and dark reactions occurred in different places. This led later to the discovery of photosystem 1 and 2.

Photodissociation

Photodissociation, photolysis, or photodecomposition is a chemical reaction in which a chemical compound is broken down by photons. It is defined as the interaction of one or more photons with one target molecule. Photodissociation is not limited to visible light. Any photon with sufficient energy can affect the chemical bonds of a chemical compound. Since a photon's energy is inversely proportional to its wavelength, electromagnetic waves with the energy of visible light or higher, such as ultraviolet light, x-rays and gamma rays are usually involved in such reactions.

Photolysis in Photosynthesis

Photolysis is part of the light-dependent reactions of photosynthesis. The general reaction of photosynthetic photolysis can be given as

$$H_2A + 2 \text{ photons (light)} \rightarrow 2\ e^- + 2\ H^+ + A$$

The chemical nature of "A" depends on the type of organism. In purple sulfur bacteria, hydrogen sulfide (H_2S) is oxidized to sulfur (S). In oxygenic photosynthesis, water (H_2O) serves as a substrate for photolysis resulting in the generation of diatomic oxygen (O_2). This is the process which returns oxygen to earth's atmosphere. Photolysis of water occurs in the thylakoids of cyanobacteria and the chloroplasts of green algae and plants.

Energy Transfer Models

The conventional, semi-classical, model describes the photosynthetic energy transfer process as one in which excitation energy hops from light-capturing pigment molecules to reaction centre molecules step-by-step down the molecular energy ladder.

The effectiveness of photons of different wavelengths depends on the absorption spectra of the photosynthetic pigments in the organism. Chlorophylls absorb light in the violet-blue and red parts of the spectrum, while accessory pigments capture other wavelengths as well. The phycobilins of red algae absorb blue-green light which penetrates deeper into water than red light, enabling them to photosynthesize in deep waters. Each absorbed photon causes the formation of an exciton (an electron excited to a higher energy state) in the pigment molecule.

The energy of the exciton is transferred to a chlorophyll molecule (P680, where P stands for pigment and 680 for its absorption maximum at 680 nm) in the reaction centre of photosystem II via resonance energy transfer. P680 can also directly absorb a photon at a suitable wavelength.

Photolysis during photosynthesis occurs in a series of light-driven oxidation events. The energized electron (exciton) of P680 is captured by a primary electron acceptor of the photosynthetic electron transfer chain and thus exits photosystem II. In order to repeat the reaction, the electron in the reaction centre needs to be replenished. This occurs by oxidation of water in the case of oxygenic photosynthesis. The electron-deficient reaction centre of photosystem II (P680*) is the strongest biological oxidizing agent yet discovered, which allows it to break apart molecules as stable as water.

The water-splitting reaction is catalyzed by the oxygen evolving complex of photosystem II. This protein-bound inorganic complex contains four manganese ions, plus calcium and chloride ions as cofactors. Two water molecules are complexed by the manganese cluster, which then undergoes a series of four electron removals (oxidations) to replenish the reaction centre of photosystem II. At the end of this cycle, free oxygen (O_2) is generated and the hydrogen of the water molecules has been converted to four protons released into the thylakoid lumen.

These protons, as well as additional protons pumped across the thylakoid membrane coupled with the electron transfer chain, form a proton gradient across the membrane that drives photophosphorylation and thus the generation of chemical energy in the form of adenosine triphosphate (ATP). The electrons reach the P700 reaction centre of photosystem I where they are energized again by light. They are passed down another electron transfer chain and finally combine with the coenzyme $NADP^+$ and protons outside the thylakoids to NADPH. Thus, the net oxidation reaction of water photolysis can be written as:

$2\ H_2O + 2\ NADP^+ + 8$ photons (light) $\rightarrow 2\ NADPH + 2\ H^+ + O_2$

The free energy change (ΔG) for this reaction is 102 kilocalories per mole. Since the energy of light at 700 nm is about 40 kilocalories per mole of photons, approximately 320 kilocalories of light energy are available for the reaction. Therefore, approximately one-third of the available light energy is captured as NADPH during photolysis and electron transfer. An equal amount of ATP is generated by the resulting proton gradient. Oxygen as a byproduct is of no further use to the reaction and thus released into the atmosphere.

Quantum Models

In 2007 a quantum model was proposed by Graham Fleming and his co-workers which includes the possibility that photosynthetic energy transfer might involve quantum oscillations, explaining its unusually high efficiency.

According to Fleming there is direct evidence that remarkably long-lived wavelike electronic quantum coherence plays an important part in energy transfer processes during photosynthesis, which can explain the extreme efficiency of the energy transfer because it enables the system to sample all the potential energy pathways, with low loss, and choose the most efficient one.

This approach has been further investigated by Gregory Scholes and his team at the University of Toronto, which in early 2010 published research results that indicate that some marine algae make use of quantum-coherent electronic energy transfer (EET) to enhance the efficiency of their energy harnessing.

Photolysis in the Atmosphere

Photolysis also occurs in the atmosphere as part of a series of reactions by which primary pollutants such as hydrocarbons and nitrogen oxides react to form secondary pollutants such as peroxyacyl nitrates.

The two most important photodissociaton reactions in the troposphere are firstly:

$O_3 + h\nu \rightarrow O_2 + O(^1D)\ \lambda < 320$ nm

which generates an excited oxygen atom which can react with water to give the hydroxyl radical:

$O(^1D) + H_2O \rightarrow 2\ ^{\cdot}OH$

The hydroxyl radical is central to atmospheric chemistry as it initiates the oxidation of hydrocarbons in the atmosphere and so acts as a detergent.

Secondly the reaction:

$NO_2 + h\nu \rightarrow NO + O$

is a key reaction in the formation of tropospheric ozone.

The formation of the ozone layer is also caused by photodissociation. Ozone in the Earth's stratosphere is created by ultraviolet light striking oxygen molecules containing two oxygen atoms (O_2), splitting them into individual oxygen atoms (atomic oxygen).

The atomic oxygen then combines with unbroken O_2 to create ozone, O_3. In addition, photolysis is the process by which CFCs are broken down in the upper atmosphere to form ozone-destroying chlorine free radicals.

Astrophysics

In astrophysics, photodissociation is one of the major processes through which molecules are broken down (but new molecules are being formed). Because of the vacuum of the interstellar medium, molecules and free radicals can exist for a long time. Photodissociation is the main path by which molecules are broken down. Photodissociation rates are important in the study of the composition of interstellar clouds in which stars are formed.

Examples of photodissociation in the interstellar medium are ($h\nu$ is the energy of a single photon of frequency ν):

$$H_2O + h\nu \rightarrow H + OH$$

$$CH_4 + h\nu \rightarrow CH_3 + H$$

Atmospheric Gamma Ray Bursts

Currently orbiting satellites detect an average of about one gamma-ray burst per day. Because gamma-ray bursts are visible to distances encompassing most of the observable universe, a volume encompassing many billions of galaxies, this suggests that gamma-ray bursts must be exceedingly rare events per galaxy.

Measuring the exact rate of Gamma Ray bursts is difficult, but for a galaxy of approximately the same size as the Milky Way, the expected rate (for long GRBs) is about one burst every 100,000 to 1,000,000 years. Only a few percent of these would be beamed towards Earth. Estimates of rates of short GRBs are even more uncertain because of the unknown beaming fraction, but are probably comparable.

A gamma-ray burst in the Milky Way, if close enough to Earth and beamed towards it, could have significant effects on the biosphere. The absorption of radiation in the atmosphere would cause photodissociation of nitrogen, generating nitric oxide that would act as a catalyst to destroy ozone.

The atmospheric photodissociation

- N_2 -> 2N
- O_2 -> 2O
- CO_2 -> C + 2O
- H_2O -> 2H + O
- $2NH_3$ -> $3H_2$ + N_2

would yield

- NO_2 (consumes up to 400 Ozone molecules)
- CH_2 (nominal)
- CH_4 (nominal)
- CO_2

(incomplete)

According to a 2004 study, a GRB at a distance of about a kiloparsec could destroy up to half of Earth's ozone layer; the direct UV irradiation from the burst combined with additional solar UV radiation passing through the diminished ozone layer could then have potentially significant impacts on the food chain and potentially trigger a mass extinction. The authors estimate that one such burst is expected per billion years, and hypothesize that the Ordovician-Silurian extinction event could have been the result of such a burst.

There are strong indications that long gamma-ray bursts preferentially or exclusively occur in regions of low metallicity. Because the Milky Way has been metal-rich since before the Earth formed, this effect may diminish or even eliminate the possibility that a long gamma-ray burst has occurred within the Milky Way within the past billion years. No such metallicity biases are known for short gamma-ray bursts. Thus, depending on their local rate and beaming properties, the possibility for a nearby event to have had a large impact on Earth at some point in geological time may still be significant.

Multiple Photon Dissociation

Single photons in the infrared spectral range usually are not energetic enough for direct photodissociation of molecules. However, after absorption of multiple infrared photons a molecule may gain internal energy to overcome its barrier for dissociation. Multiple photon dissociation (MPD, IRMPD with infrared radiation) can be achieved by applying high power lasers, e.g. a carbon dioxide laser, or a free electron laser, or by long interaction times of the molecule with the radiation field without the possibility for rapid cooling, e.g. by collisions. The latter method allows even for MPD induced by black body radiation, a technique called Blackbody infrared radiative dissociation (BIRD).

Oxygen Evolution

Oxygen evolution is the process of generating molecular oxygen through chemical reaction. Mechanisms of oxygen evolution include the oxidation of water during oxygenic photosynthesis, electrolysis of water into oxygen and hydrogen, and electrocatalytic oxygen evolution from oxides and oxoacids.

Oxygen Evolution in Nature

Photosynthetic oxygen evolution is the fundamental process by which breathable oxygen is generated in earth's biosphere. The reaction is part of the light-dependent reactions of photosynthesis in cyanobacteria and the chloroplasts of green algae and plants. It utilises the energy of light to split a water molecule into its protons and electrons for photosynthesis. Free oxygen is generated as a by-product of this reaction, and is released into the atmosphere.

Biochemical Reaction

Photosynthetic oxygen evolution occurs via the light-dependent oxidation of water to molecular oxygen and can be written as the following simplified chemical reaction:

$$2H_2O \rightarrow 4e^- + 4H^+ + O_2$$

The reaction requires the energy of four photons. The electrons from the oxidized water molecules replace electrons in the P_{680} component of photosystem II that have been removed into an electron transport chain via light-dependent excitation and resonance energy transfer onto plastoquinone. Photosytem II, therefore, has also been referred to as water-plastoquinone oxido-reductase.

The protons are released into the thylakoid lumen, thus contributing to the generation of a proton gradient across the thylakoid membrane. This proton gradient is the driving force for ATP synthesis via photophosphorylation and coupling the absorption of light energy and oxidation of water to the creation of chemical energy during photosynthesis.

Oxygen-Evolving Complex

Water oxidation is catalyzed by a manganese-containing cofactor contained in photosystem II known as the oxygen-evolving complex (OEC) or water-splitting complex. Manganese is an important cofactor, and calcium and chloride are also required for the reaction to occur.

X-ray crystallographic data have been used to propose a structure and mechanism of action for the oxygen-evolving complex and its manganese cluster. Based on structural and spectroscopic experiments, oxygen evolution involves a core three-plus-one cluster of three manganese ions and one calcium ion, with one additional manganese, which are oxidized via intermediate states called *S-states*. The O-O bond of molecular oxygen is formed between manganese-ligated oxygen atoms at the most oxidized, or S4, state.

History of Discovery

It was not until the end of the 18th century that Joseph Priestley discovered by accident the ability of plants to "restore" air that had been "injured" by the burning of a candle. He followed up on the experiment by showing that air "restored" by vegetation was *"not at all inconvenient to a mouse."* He was later awarded a medal for his discoveries that: *"...no vegetable grows in vain... but cleanses and purifies our atmosphere."* Priestley's experiments were followed up by Jan Ingenhousz, a Dutch physician, who showed that "restoration" of air only worked in the presence of light and green plant parts.

Ingenhousz suggested in 1796 that CO_2 (carbon dioxide) is split during photosynthesis to release oxygen, while the carbon combined with water to form carbohydrates. While this hypothesis was attractive and reasonable and thus widely accepted for a long time, it was later proven incorrect. Graduate student C.B. Van Niel at Stanford University found that purple sulfur bacteria reduce carbon to carbohydrates, but accumulate sulfur instead of releasing oxygen.

He boldly proposed that, in analogy to the sulfur bacteria's forming elemental sulfur from H_2S (hydrogen sulfide), plants would form oxygen from H_2O (water). In 1937, this hypothesis was corroborated by the discovery that plants are capable of producing oxygen in the absence of CO_2.

This discovery was made by Robin Hill, and subsequently the light-driven release of oxygen in the absence of CO_2 was called the *Hill reaction*. Our current knowledge of the mechanism of oxygen evolution during photosynthesis was further established in experiments tracing isotopes of oxygen from water to oxygen gas.

Technological Oxygen Evolution

Oxygen evolution occurs as a byproduct of hydrogen production via electrolysis of water. While oxygen production is not the main focus of industrial applications of water electrolysis, it becomes essential for life support systems in situations that require the generation of oxygen for air revitalization. Human exploration of regions that lack breathable oxygen, such as the deep sea or outer space, requires means of reliably generating oxygen apart from earth's atmosphere. Submarines and spacecraft utilise either an electrolytic mechanism (water or solid oxide electrolysis) or chemical oxygen generators as part of their life support equipment.

Carbon Fixation

Carbon fixation is the conversion of inorganic carbon (carbon dioxide) to organic compounds by living organisms. The most prominent example is photosynthesis, although chemosynthesis is another form of carbon fixation that can take place in the absence of sunlight. Organisms that grow by fixing carbon are called autotrophs. Autotrophs include photoautotrophs, which synthesize organic compounds using the energy of sunlight, and lithoautotrophs, which synthesize organic compounds using the energy of inorganic oxidation. Heterotrophs are organisms that grow using the carbon fixed by autotrophs.

The organic compounds are used by heterotrophs to produce energy and to build body structures. "Fixed carbon", "reduced carbon", and "organic carbon" are equivalent terms for various organic compounds.

Net vs Gross CO_2 Fixation

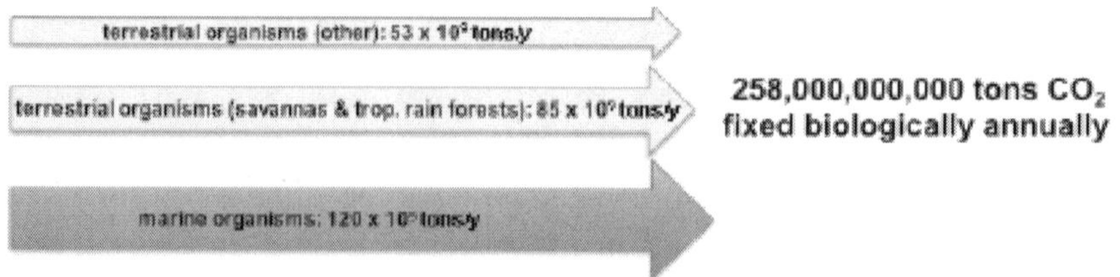

Figure: *Graphic showing net annual amounts of CO_2 fixation by land and sea-based organisms.*

It is estimated that approximately 258 billion tons of carbon dioxide are converted by photosynthesis annually. The majority of the fixation occurs in marine environments, especially areas of high nutrients. The gross amount of carbon dioxide fixed is much larger since approximately 40% is consumed by respiration in the evenings following each day of photosynthesis. Given the scale of this process, it is understandable that RuBisCO is the most abundant protein on earth.

Overview of Pathways

Six autotrophic carbon fixation pathways are known as of 2011. The Calvin cycle fixes carbon in the chloroplasts of plants and algae, and in the cyanobacteria. It also fixes carbon in the anoxygenic photosynthetic proteobacteria called purple bacteria, and in some non-phototrophic proteobacteria.

Oxygenic Photosynthesis

In photosynthesis, energy from sunlight drives the carbon fixation pathway. *Oxygenic* photosynthesis is used by the primary producers—plants, algae, and cyanobacteria. They contain the pigment chlorophyll, and use the Calvin cycle to fix carbon autotrophically. The process works like this:

$$2H_2O \rightarrow 4e^- + 4H^+ + O_2$$

$$CO_2 + 4e^- + 4H^+ \rightarrow CH_2O + H_2O$$

In the first step, water is dissociated into electrons, protons, and free oxygen. This allows the use of water, one of the most abundant substances on Earth, as an electron donor—as a source of reducing power. The release of free oxygen is a side-effect of enormous consequence. The first step uses the energy of sunlight to oxidize water to O_2, and, ultimately, to produce ATP

$$ADP + P_i \rightleftharpoons ATP + H_2O$$

and the reductant, NADPH

$$NADP^+ + 2e^- + 2H^+ \rightleftharpoons NADPH + H^+$$

The second step, called the Calvin cycle, the actual fixation of carbon dioxide is carried out. This process consumes ATP and NADPH. The Calvin cycle in plants accounts for the preponderance of carbon fixation on land. In algae and cyanobacteria, it accounts for the preponderance of carbon fixation in the oceans. The Calvin cycle converts carbon dioxide into sugar, as triose phosphate (TP), which is glyceraldehyde 3-phosphate (GAP) together with dihydroxyacetone phosphate (DHAP):

$$3\ CO_2 + 12\ e^- + 12\ H^+ + P_i \rightarrow TP + 4\ H_2O$$

An alternative perspective accounts for NADPH (source of e^-) and ATP:

$3\ CO_2 + 6\ NADPH + 6\ H^+ + 9\ ATP + 5\ H_2O \rightarrow TP + 6\ NADP^+ + 9\ ADP + 8\ P_i$

The formula for inorganic phosphate (P_i) is $HOPO_3^{2-} + 2H^+$. Formulas for triose and TP are $C_2H_3O_2\text{-}CH_2OH$ and $C_2H_3O_2\text{-}CH_2OPO_3^{2-} + 2H^+$

Evolutionary Considerations

Somewhere between 3.5 and 2.3 billion years ago, cyanobacteria evolved oxygenic photosynthesis.

Carbon Concentrating Mechanisms

Many photosynthetic organisms have acquired inorganic carbon concentrating mechanisms (CCM), which increase the concentration of carbon dioxide available to the initial carboxylase of the Calvin cycle, the enzyme RuBisCO. The benefits of CCM include increased tolerance to low external concentrations of inorganic carbon, and reduced loses to photorespiration. CCM can make plants more tolerant of heat and water stress.

Carbon concentrating mechanisms use the enzyme carbonic anhydrase (CA), which catalyze both the dehydration of bicarbonate to carbon dioxide and the hydration of carbon dioxide to bicarbonate

$HCO_3^- + H^+ \rightleftharpoons CO_2 + H_2O$

Lipid membranes are much less permeable to bicarbonate than to carbon dioxide. To capture inorganic carbon more effectively, some plants have adapted the anaplerotic reaction

$HCO_3^- + H^+ + PEP \rightarrow OAA + P_i$

catalyzed by PEP carboxylase (PEPC), to carboxylate phosphoenol-pyruvate (PEP) to oxaloacetate (OAA) which is a C_4 dicarboxylic acid.

CAM Plants

CAM plants that use Crassulacean acid metabolism as an adaptation for arid conditions. CO_2 enters through the stomata during the night and is converted into the 4-carbon compound, malic acid, which releases CO_2 for use in the Calvin cycle during the day, when the stomata are closed. The jade plant (*Crassula ovata*) and cacti are typical of CAM plants. Sixteen thousand species of plants use CAM. These plants have a carbon isotope signature of -20 to -10 ‰.

C_4 Plants

C_4 plants preface the Calvin cycle with reactions that incorporate CO_2 into one of the 4-carbon compounds, malic acid or aspartic acid.

C_4 plants have a distinctive internal leaf anatomy. Tropical grasses, such as sugar cane and maize are C_4 plants, but there are many broadleaf plants that are C_4. Overall, 7600 species of terrestrial plants use C_4 carbon fixation, representing around 3% of all species. These plants have a carbon isotope signature of -16 to -10 ‰.

C_3 Plants

The large majority of plants are C_3 plants. They are so-called to distinguish them from the CAM and C_4 plants, and because the carboxylation products of the Calvin cycle are 3-carbon compounds. They lack C_4 dicarboxylic acid cycles, and therefore have higher carbon dioxide compensation points than CAM or C_4 plants. C_3 plants have a carbon isotope signature of -24 to -33‰.

Other Autotrophic Pathways

Of the five other autotrophic pathways, two are known only in bacteria, two only in archaea, and one in both bacteria and archaea.

Reductive Citric Acid Cycle

The reductive citric acid cycle is the oxidative citric acid cycle run in reverse. It has been found in anaerobic and microaerobic bacteria. It was proposed in 1966 by Evans, Buchanan and Arnon who were working with the anoxygenic photosynthetic green sulfur bacterium that they called *Chlorobium thiosulfatophilum*. The reductive citric acid cycle is sometimes called the Arnon-Buchanan cycle.

Reductive Acetyl CoA Pathway

The reductive acetyl CoA pathway operated in strictly anaerobic bacteria and archaea. The pathway was proposed in 1965 by Ljungdahl and Wood.

They were working with the gram-positive acetic acid producing bacterium *Clostridium thermoaceticum*, which is now named *Moorella thermoacetica*. The reductive acetyl CoA pathway is sometimes called the Wood-Ljungdahl pathway.

3-Hydroxypropionate and Two Related Cycles

The 3-hydroxypropionate cycle is utilised *only* by green nonsulfur bacteria. It was proposed in 2002 for the anoxygenic photosynthetic *Chloroflexus aurantiacus*. None of the enzymes that participate in the 3-hydroxypropionate cycle are especially oxygen sensitive.

A variant of the 3-hydroxypropionate pathway was found to operated in aerobic extreme thermoacidophile archaeon *Metallosphaera sedula*.

This pathway, called the 3-hydroxypropionate/4-hydroxybutyrate cycle . And yet another variant of the 3-hydroxypropionate pathway is the dicarboxylate/4-hydroxybutyrate cycle. It was discovered in anaerobic archaea. It was proposed in 2008 for the hyperthermophile archeon *Ignicoccus hospitalis.*

Chemosynthesis

Chemosynthesis is carbon fixation driven by the oxidation of inorganic substances (e.g., hydrogen gas or hydrogen sulfide). Sulfur- and hydrogen-oxidizing bacteria often use the Calvin cycle or the reductive citric acid cycle.

Non-Autotrophic Pathways

Although almost all heterotrophs cannot synthesize complete organic molecules from carbon dioxide, some carbon dioxide is incorporated in their metabolism. Notably pyruvate carboxylase consumes carbon dioxide (as bicarbonate ions) as part of gluconeogenesis, and carbon dioxide is consumed in various anaplerotic reactions.

Carbon Isotope Discrimination

Some carboxylases, particularly RuBisCO, preferentially bind the lighter carbon stable isotope carbon-12 over the heavier carbon-13. This is known as carbon isotope discrimination and results in carbon-12 to carbon-13 ratios in the plant that are lower than in the free air. Measurement of this ratio is important in the evaluation of water use efficiency in plants, and also in assessing the possible or likely sources of carbon in global carbon cycle studies.

Carbon Concentrating Mechanisms

On land: In hot and dry conditions, plants close their stomata to prevent the loss of water. Under these conditions, CO_2 will decrease, and oxygen gas, produced by the light reactions of photosynthesis, will decrease in the stem, not leaves, causing an increase of photorespiration by the oxygenase activity of ribulose-1,5-bisphosphate carboxylase/ oxygenase and decrease in carbon fixation. Some plants have evolved mechanisms to increase the CO_2 concentration in the leaves under these conditions.

C_4 plants chemically fix carbon dioxide in the cells of the mesophyll by adding it to the three-carbon molecule phosphoenolpyruvate (PEP), a reaction catalyzed by an enzyme called PEP carboxylase, creating the four-carbon organic acid oxaloacetic acid. Oxaloacetic acid or malate synthesized by this process is then translocated to specialised bundle sheath cells where the enzyme RuBisCO and other Calvin cycle enzymes are located, and where CO_2 released by decarboxylation of the four-carbon acids is then fixed by RuBisCO activity to the three-carbon sugar 3-phosphoglyceric acids.

The physical separation of RuBisCO from the oxygen-generating light reactions reduces photorespiration and increases CO_2 fixation and, thus, photosynthetic capacity of the leaf. C_4 plants can produce more sugar than C_3 plants in conditions of high light and temperature. Many important crop plants are C_4 plants, including maize, sorghum, sugarcane, and millet. Plants that do not use PEP-carboxylase in carbon fixation are called C_3 plants because the primary carboxylation reaction, catalyzed by RuBisCO, produces the three-carbon sugar 3-phosphoglyceric acids directly in the Calvin-Benson cycle. Over 90% of plants use C_3 carbon fixation, compared to 3% that use C_4 carbon fixation.; however, the fact that C_4 has evolved in over 60 plant lineages makes it a striking example of convergent evolution.

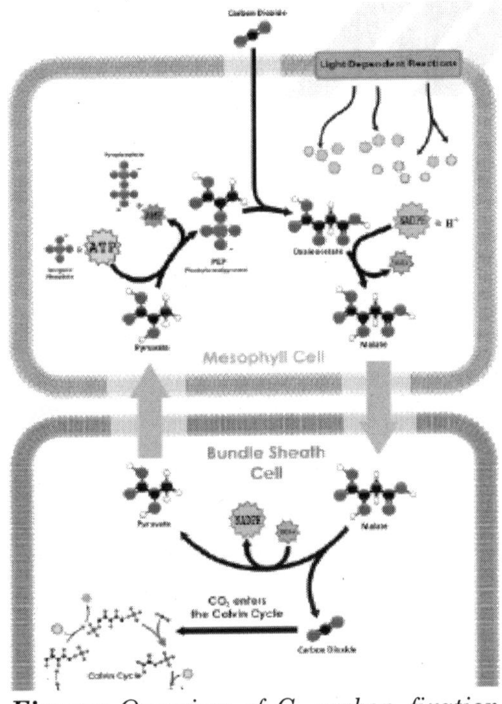

***Figure:** Overview of C_4 carbon fixation*

Xerophytes, such as cacti and most succulents, also use PEP carboxylase to capture carbon dioxide in a process called Crassulacean acid metabolism (CAM). In contrast to C_4 metabolism, which *physically* separates the CO_2 fixation to PEP from the Calvin cycle, CAM *temporally* separates these two processes. CAM plants have a different leaf anatomy from C_3 plants, and fix the CO_2 at night, when their stomata are open. CAM plants store the CO_2 mostly in the form of malic acid via carboxylation of phosphoenolpyruvate to oxaloacetate, which is then reduced to malate. Decarboxylation of malate during the day releases CO_2 inside the leaves, thus allowing carbon fixation to 3-phosphoglycerate by RuBisCO. Sixteen thousand species of plants use CAM.

In Water

Cyanobacteria possess carboxysomes, which increase the concentration of CO_2 around RuBisCO to increase the rate of photosynthesis. An enzyme, carbonic anhydrase, located within the carboxysome releases CO_2 from the dissolved hydrocarbonate ions (HCO_3^-). Before the CO_2 diffuses out it is quickly sponged up by RuBisCO, which is concentrated within the carboxysomes. HCO_3^- ions are made from CO_2 outside the cell by another carbonic anhydrase and are actively pumped into the cell by a membrane protein.

They cannot cross the membrane as they are charged, and within the cytosol they turn back into CO_2 very slowly without the help of carbonic anhydrase. This causes the HCO_3^- ions to accumulate within the cell from where they diffuse into the carboxysomes. Pyrenoids in algae and hornworts also act to concentrate CO_2 around rubisco.

Order and Kinetics

The overall process of photosynthesis takes place in four stages:

Stage	*Description*	*Time scale*
1	Energy transfer in antenna chlorophyll (thylakoid membranes)	femtosecond to picosecond
2	Transfer of electrons in photochemical reactions (thylakoid membranes)	picosecond to nanosecond
3	Electron transport chain and ATP synthesis (thylakoid membranes)	microsecond to millisecond
4	Carbon fixation and export of stable products	millisecond to second

Photosynthetic Efficiency

The photosynthetic efficiency is the fraction of light energy converted into chemical energy during photosynthesis in plants and algae. Photosynthesis can be described by the simplified chemical reaction

$$6H_2O + 6CO_2 + \text{energy} \rightarrow C_6H_{12}O_6 + 6O_2$$

where $C_6H_{12}O_6$ is glucose (which is subsequently transformed into other sugars, cellulose, lignin, and so forth). The value of the photosynthetic efficiency is dependent on how light energy is defined – it depends on whether we count only the light that is absorbed, and on what kind of light is used. It takes eight (or perhaps 10 or more) photons to utilise one molecule of CO_2. The Gibbs free energy for converting a mole of CO_2 to glucose is 114 kcal, whereas eight moles of photons of wavelength 600 nm contains 381 kcal, giving a nominal efficiency of 30%.

However, photosynthesis can occur with light up to wavelength 720 nm so long as there is also light at wavelengths below 680 nm to keep Photosystem II operating. Using longer wavelengths means less light energy is needed for the same number of photons and therefore for the same amount of photosynthesis. For actual sunlight, where only 45% of the light is in the photosynthetically active wavelength range, the theoretical maximum efficiency of solar energy conversion is approximately 11%. In actuality, however, plants do not absorb all incoming sunlight (due to reflection, respiration requirements of photosynthesis and the need for optimal solar radiation levels) and do not convert all harvested energy into biomass, which results in an overall photosynthetic efficiency of 3 to 6% of total solar radiation. If photosynthesis is inefficient, excess light energy must be dissipated to avoid damaging the photosynthetic apparatus. Energy can be dissipated as heat (non-photochemical quenching), or emitted as chlorophyll fluorescence.

Typical Efficiencies

Plants:

Quoted values sunlight-to-biomass efficiency

Plant	*Efficiency*
Plants, typical	0.1% 0.2–2%
Typical crop plants	1–2%
Sugarcane	7–8% peak

The following is a breakdown of the energetics of the photosynthesis process from *Photosynthesis* by Hall and Rao:

Starting with the solar spectrum falling on a leaf, 47% lost due to photons outside the 400–700 nm active range (chlorophyll utilises photons between 400 and 700 nm, extracting the energy of one 700 nm photon from each one) 30% of the in-band photons are lost due to incomplete absorption or photons hitting components other than chloroplasts 24% of the absorbed photon energy is lost due to degrading short wavelength photons to the 700 nm energy level 68% of the utilised energy is lost in conversion into d-glucose 35–45% of the glucose is consumed by the leaf in the processes of dark and photo respiration

Stated another way: 100% sunlight → non-bioavailable photons waste is 47%, leaving 53% (in the 400–700 nm range) → 30% of photons are lost due to incomplete absorption, leaving 37% (absorbed photon energy) → 24% is lost due to wavelength-mismatch degradation to 700 nm energy, leaving 28.2% (sunlight energy collected by chlorophyl) → 32% efficient conversion of ATP and NADPH to d-glucose, leaving 9% (collected as sugar) → 35–40% of sugar is recycled/ consumed by the leaf in dark and photo-respiration, leaving 5.4% net leaf efficiency.

Many plants lose much of the remaining energy on growing roots. Most crop plants store ~0.25% to 0.5% of the sunlight in the product (corn kernels, potato starch, etc.). Sugar cane is exceptional in several ways, yielding peak storage efficiencies of ~8%.

Figure: *Measuring the photosynthetic efficiency of wheat in the field using an LCpro-SD*

Photosynthesis increases linearly with light intensity at low intensity, but at higher intensity this is no longer the case. Above about 10,000 lux or ~100 watts/square metre the rate no longer increases. Thus, most plants can only utilise ~10% of full mid-day sunlight intensity. This dramatically reduces average achieved photosynthetic efficiency in fields compared to peak laboratory results. However, real plants (as opposed to laboratory test samples) have lots of redundant, randomly oriented leaves. This helps to keep the average illumination of each leaf well below the mid-day peak enabling the plant to achieve a result closer to the expected laboratory test results using limited illumination. Only if the light intensity is above a plant specific value, called the compensation point the plant assimilates more carbon and releases more oxygen by photosynthesis than it consumes by cellular respiration for its own current energy demand.

Photosynthesis measurement systems are not designed to directly measure the amount of light absorbed by the leaf. Nevertheless, the light response curves that the class produces do allow comparisons in photosynthetic efficiency between plants.

Algae and Other Monocellular Organisms

From a 2010 study by the University of Maryland, photosynthesizing Cyanobacteria have been shown to be a significant species in the global carbon cycle, accounting for 20–30% of Earth's photosynthetic productivity and convert solar energy into biomass-stored chemical energy at the rate of ~450 TW.

Worldwide Figures

According to the cyanobacteria study above, this means the total photosynthetic productivity of earth is between ~1500–2250 TW, or 47,300–71,000 exajoules per year. Using this source's figure of 178,000 TW of solar energy hitting the Earth's surface, the total photosynthetic efficiency of the planet is 0.84% to 1.26%.

Efficiencies of Various Biofuel Crops

Popular choices for plant biofuels include: oil palm, soybean, castor oil, sunflower oil, safflower oil, corn ethanol, and sugar cane ethanol.

An analysis of a proposed Hawaiian oil palm plantation claimed to yield 600 gallons of biodiesel per acre per year. That comes to 2835 watts per acre or 0.7 W/m^2. Typical insolation in Hawaii is around 5.5 $kWh/(m^2day)$ or 230 watts. For this particular oil palm plantation, if it delivered the claimed 600 gallons of biodiesel per acre per year, would be converting 0.3% of the incident solar energy to chemical fuel. Total photosynthetic efficiency would include more than just the biodiesel oil, so this 0.3% number is something of a lower bound.

Contrast this with a typical photo-voltaic installation, which would produce an average of roughly 22 W/m^2 (roughly 10% of the average insolation), throughout the year. Most crop plants store ~0.25% to 0.5% of the sunlight in the product (corn kernels, potato starch, etc.), sugar cane is exceptional in several ways to yield peak storage efficiencies of ~8%.

Ethanol fuel in Brazil has a calculation that results in: "Per hectare per year, the biomass produced corresponds to 0.27 TJ. This is equivalent to 0.86 W/m^2. Assuming an average insolation of 225 W/m^2, the photosynthetic efficiency of sugar cane is 0.38%." Sucrose accounts for little more than 30% of the chemical energy stored in the mature plant; 35% is in the leaves and stem tips, which are left in the fields during harvest, and 35% are in the fibrous material (bagasse) left over from pressing.

C3 vs. C4 and CAM Plants

C3 plants use the Calvin cycle to fix carbon. C4 plants use a modified Calvin cycle in which they separate Ribulose-1,5-bisphosphate carboxylase oxygenase (RuBisCO) from atmospheric oxygen, fixing carbon in their mesophyll cells and using oxaloacetate and malate to ferry the fixed carbon to RuBisCO and the rest of the Calvin cycle enzymes isolated in the bundle-sheath cells.

The intermediate compounds both contain four carbon atoms, which gives C4. In Crassulacean acid metabolism (CAM), time isolates functioning RuBisCo (and the other Calvin cycle enzymes) from high oxygen concentrations produced by photosynthesis, in that O_2 is evolved during the day, and allowed to dissipate then, while at night atmospheric CO_2 is taken up and stored as malic or other acids. During the day, CAM plants close stomata and use stored acids as carbon sources for sugar, etc. production.

The C3 pathway requires 18 ATP for the synthesis of one molecule of glucose while the C4 pathway requires 30 ATP. C4 is an evolutionary advancement over the simpler C3 cycle which operates in most plants. Corn, sugar cane, and sorghum are C4 plants. These plants are economically important in part because of their relatively high photosynthetic efficiencies compared to many other crops. Pineapple is a CAM plant.

Evolution of Photosynthesis

The evolution of photosynthesis refers to the origin and subsequent evolution of photosynthesis, the process by which light energy from the sun is used to synthesize sugars from carbon dioxide, releasing oxygen as a waste product.

The first photosynthetic organisms probably evolved early in the evolutionary history of life and most likely used reducing agents such as hydrogen or hydrogen sulfide as sources of electrons, rather than water. There are three major metabolic pathways by which photosynthesis is carried out: C_3 photosynthesis, C_4 photosynthesis, and CAM photosynthesis. C_3 photosynthesis is the oldest and most common form.

Origin

The biochemical capacity to use water as the source for electrons in photosynthesis evolved once, in a common ancestor of extant cyanobacteria. The geological record indicates that this transforming event took place early in Earth's history, at least 2450–2320 million years ago (Ma), and, it is speculated, much earlier. Available evidence from geobiological studies of Archean (>2500 Ma) sedimentary rocks indicates that life existed 3500 Ma, but the question of when oxygenic photosynthesis evolved is still unanswered.

A clear paleontological window on cyanobacterial evolution opened about 2000 Ma, revealing an already-diverse biota of blue-greens. Cyanobacteria remained principal primary producers throughout the Proterozoic Eon (2500–543 Ma), in part because the redox structure of the oceans favoured photoautotrophs capable of nitrogen fixation. Green algae joined blue-greens as major primary producers on continental shelves near the end of the Proterozoic, but only with the Mesozoic (251–65 Ma) radiations of dinoflagellates, coccolithophorids, and diatoms did primary production in marine shelf waters take modern form.

Cyanobacteria remain critical to marine ecosystems as primary producers in oceanic gyres, as agents of biological nitrogen fixation, and, in modified form, as the plastids of marine algae.

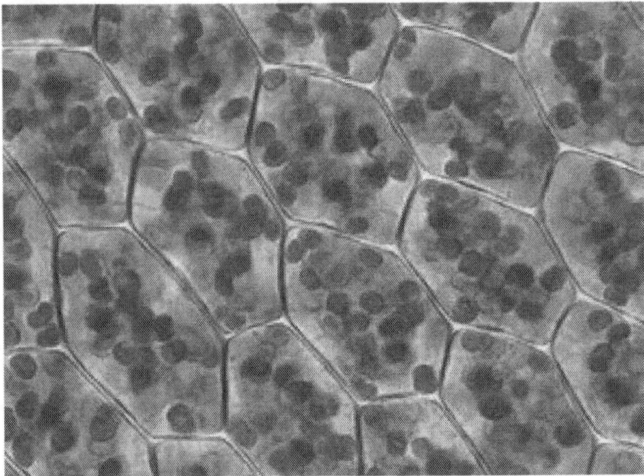

Figure: *Plant cells with visible chloroplasts (from a moss,* Plagiomnium affine*)*

Early photosynthetic systems, such as those from green and purple sulfur and green and purple nonsulfur bacteria, are thought to have been anoxygenic, using various molecules as electron donors. Green and purple sulfur bacteria are thought to have used hydrogen and sulfur as an electron donor. Green nonsulfur bacteria used various amino and other organic acids. Purple nonsulfur bacteria used a variety of nonspecific organic molecules. The use of these molecules is consistent with the geological evidence that the atmosphere was highly reduced at that time.

Fossils of what are thought to be filamentous photosynthetic organisms have been dated at 3.4 billion years old.

The main source of oxygen in the atmosphere is oxygenic photosynthesis, and its first appearance is sometimes referred to as the oxygen catastrophe. Geological evidence suggests that oxygenic photosynthesis, such as that in cyanobacteria, became important during the Paleoproterozoic era around 2 billion years ago. Modern photosynthesis in plants and most photosynthetic prokaryotes is oxygenic. Oxygenic photosynthesis uses water as an electron donor, which is oxidized to molecular oxygen (O_2) in the photosynthetic reaction centre.

Symbiosis and the Origin of Chloroplasts

Several groups of animals have formed symbiotic relationships with photosynthetic algae. These are most common in corals, sponges and sea anemones. It is presumed that this is due to the particularly simple body plans and large surface areas of these animals compared to their volumes. In addition, a few marine mollusks *Elysia viridis* and *Elysia chlorotica* also maintain a symbiotic relationship with chloroplasts they capture from the algae in their diet and then store in their bodies. This allows the mollusks to survive solely by photosynthesis for several months at a time. Some of the genes from the plant cell nucleus have even been transferred to the slugs, so that the chloroplasts can be supplied with proteins that they need to survive.

An even closer form of symbiosis may explain the origin of chloroplasts. Chloroplasts have many similarities with photosynthetic bacteria, including a circular chromosome, prokaryotic-type ribosomes, and similar proteins in the photosynthetic reaction centre. The endosymbiotic theory suggests that photosynthetic bacteria were acquired (by endocytosis) by early eukaryotic cells to form the first plant cells. Therefore, chloroplasts may be photosynthetic bacteria that adapted to life inside plant cells. Like mitochondria, chloroplasts still possess their own DNA, separate from the nuclear DNA of their plant host cells and the genes in this chloroplast DNA resemble those in cyanobacteria. DNA in chloroplasts codes for redox proteins such as photosynthetic reaction centres. The CoRR Hypothesis proposes that this Co-location is required for Redox Regulation.

A 2010 study by researchers at Tel Aviv University discovered that the Oriental hornet (*Vespa orientalis*) converts sunlight into electric power using a pigment called xanthopterin. This is the first scientific evidence of a member of the animal kingdom engaging in photosynthesis.

Evolution of Photosynthetic Pathways

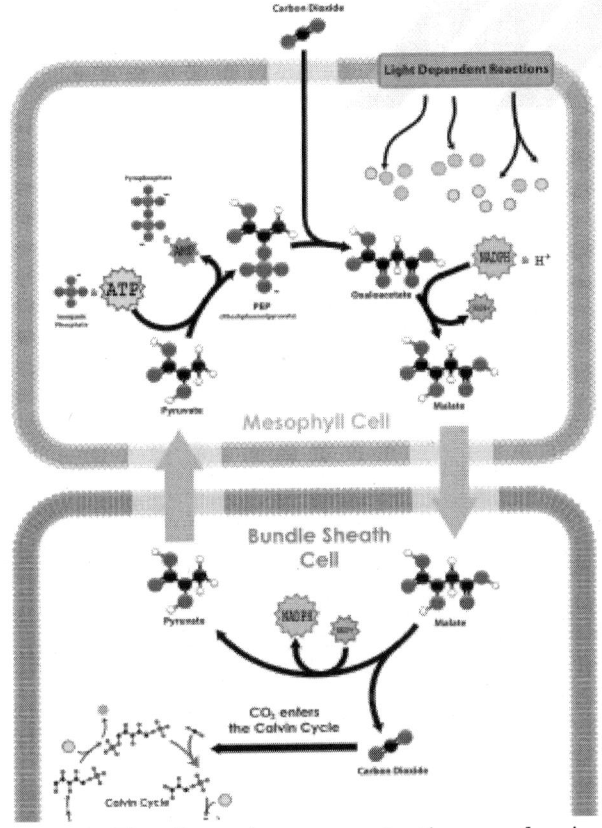

Figure: *The C_4 carbon concentrating mechanism*

Photosynthesis is not quite as simple as adding water to CO_2 to produce sugars and oxygen. A complex chemical pathway is involved, facilitated along the way by a range of enzymes and co-enzymes. The enzyme RuBisCO is responsible for "fixing" CO_2 – that is, it attaches it to a carbon-based molecule to form a sugar, which can be used by the plant, releasing an oxygen molecule along the way. However, the enzyme is notoriously inefficient, and just as effectively will also fix oxygen instead of CO_2 in a process called photorespiration. This is energetically costly as the plant has to use energy to turn the products of photorespiration back into a form that can react with CO_2.

Concentrating Carbon

The C_4 metabolic pathway is a valuable recent evolutionary innovation in plants, involving a complex set of adaptive changes to physiology and gene expression patterns. About 7600 species of plants use C_4 carbon fixation, which represents about 3% of all terrestrial species of plants. All these 7600 species are angiosperms.

C_4 plants evolved carbon concentrating mechanisms. These work by increasing the concentration of CO_2 around RuBisCO, thereby facilitating photosynthesis and decreasing photorespiration. The process of concentrating CO_2 around RuBisCO requires more energy than allowing gases to diffuse, but under certain conditions – i.e. warm temperatures (>25°C), low CO_2 concentrations, or high oxygen concentrations – pays off in terms of the decreased loss of sugars through photorespiration.

One type of C_4 metabolism employs a so-called Kranz anatomy. This transports CO_2 through an outer mesophyll layer, via a range of organic molecules, to the central bundle sheath cells, where the CO_2 is released. In this way, CO_2 is concentrated near the site of RuBisCO operation. Because RuBisCO is operating in an environment with much more CO_2 than it otherwise would be, it performs more efficiently.

Figure: *CAM is named after the family Crassulaceae, to which the jade plant belongs. Another example of a CAM plant is the pineapple.*

A second mechanism, CAM photosynthesis, is a carbon fixation pathway that evolved in some plants as an adaptation to arid conditions. The most important benefit of CAM to the plant is the ability to leave most leaf stomata closed during the day. This reduces water loss due to evapotranspiration. The stomata open at night to collect CO_2, which is stored as the four-carbon acid malate, and then used during photosynthesis during the day. The pre-collected CO_2 is concentrated around the enzyme RuBisCO, increasing photosynthetic efficiency. More CO_2 is then harvested from the atmosphere when stomata open, during the cool, moist nights, reducing water loss.

CAM has evolved convergently many times. It occurs in 16,000 species (about 7% of plants), belonging to over 300 genera and around 40 families, but this is thought to be a considerable underestimate. It is found in quillworts (relatives of club mosses), in ferns, and in gymnosperms, but the great majority of plants using CAM are angiosperms (flowering plants).

Evolutionary Record

These two pathways, with the same effect on RuBisCO, evolved a number of times independently – indeed, C_4 alone arose 62 times in 18 different plant families. A number of 'pre-adaptations' seem to have paved the way for C4, leading to its clustering in certain clades: it has most frequently been innovated in plants that already had features such as extensive vascular bundle sheath tissue. Many potential evolutionary pathways resulting in the C_4 phenotype are possible and have been characterised using Bayesian inference, confirming that non-photosynthetic adaptations often provide evolutionary stepping stones for the further evolution of C_4.

The C_4 construction is most famously used by a subset of grasses, while CAM is employed by many succulents and cacti. The trait appears to have emerged during the Oligocene, around 25 to 32 million years ago; however, they did not become ecologically significant until the Miocene, 6 to 7 million years ago. Remarkably, some charcoalified fossils preserve tissue organised into the Kranz anatomy, with intact bundle sheath cells, allowing the presence C_4 metabolism to be identified without doubt at this time. Isotopic markers are used to deduce their distribution and significance.

C_3 plants preferentially use the lighter of two isotopes of carbon in the atmosphere, ^{12}C, which is more readily involved in the chemical pathways involved in its fixation. Because C_4 metabolism involves a further chemical step, this effect is accentuated. Plant material can be analysed to deduce the ratio of the heavier ^{13}C to ^{12}C. This ratio is denoted ä^{13}C. C_3 plants are on average around 14‰ (parts per thousand) lighter than the atmospheric ratio, while C_4 plants are about 28‰ lighter. The ä^{13}C of CAM plants depends on the percentage of carbon fixed at night relative to what is fixed in the day, being closer to C_3 plants if they fix most carbon in the day and closer to C_4 plants if they fix all their carbon at night.

It is troublesome procuring original fossil material in sufficient quantity to analyse the grass itself, but fortunately there is a good proxy: horses. Horses were globally widespread in the period of interest, and browsed almost exclusively on grasses. There's an old phrase in isotope palæontology, "you are what you eat (plus a little bit)" – this refers to the fact that organisms reflect the isotopic composition of whatever they eat, plus a small adjustment factor. There is a good record of horse teeth throughout the globe, and their ä^{13}C has been measured. The record shows a sharp negative inflection around 6 to 7 million years ago, during the Messinian, and this is interpreted as the rise of C_4 plants on a global scale.

When is C_4 an Advantage?

While C_4 enhances the efficiency of RuBisCO, the concentration of carbon is highly energy intensive. This means that C_4 plants only have an advantage over C_3 organisms in certain conditions: namely, high temperatures and low rainfall. C_4 plants also need high levels of sunlight to thrive. Models suggest that, without wildfires removing shade-casting trees and shrubs, there would be no space for C_4 plants. But, wildfires have occurred for 400 million years – why did C_4 take so long to arise, and then appear independently so many times? The Carboniferous period (~300 million years ago) had notoriously high oxygen levels – almost enough to allow spontaneous combustion – and very low CO_2, but there is no C_4 isotopic signature to be found. And there doesn't seem to be a sudden trigger for the Miocene rise.

During the Miocene, the atmosphere and climate were relatively stable. If anything, CO_2 increased gradually from 14 to 9 million years ago before settling down to concentrations similar to the Holocene. This suggests that it did not have a key role in invoking C_4 evolution. Grasses themselves (the group which would give rise to the most occurrences of C_4) had probably been around for 60 million years or more, so had had plenty of time to evolve C_4, which, in any case, is present in a diverse range of groups and thus evolved independently. There is a strong signal of climate change in South Asia; increasing aridity – hence increasing fire frequency and intensity – may have led to an increase in the importance of grasslands. However, this is difficult to reconcile with the North American record. It is possible that the signal is entirely biological, forced by the fire- (and elephant?)-driven acceleration of grass evolution – which, both by increasing weathering and incorporating more carbon into sediments, reduced atmospheric CO_2 levels. Finally, there is evidence that the onset of C_4 from 9 to 7 million years ago is a biased signal, which only holds true for North America, from where most samples originate; emerging evidence suggests that grasslands evolved to a dominant state at least 15Ma earlier in South America.

Photorespiration

Photorespiration (also known as the oxidative photosynthetic carbon cycle, or C2 photosynthesis) is a process in plant metabolism which attempts to ameliorate the consequences of a wasteful oxygenation reaction by the enzyme RuBisCO. The desired reaction is the addition of carbon dioxide to RuBP (carboxylation), a key step in the Calvin-Benson cycle, however approximately 25% of reactions by RuBisCO instead add oxygen to RuBP (oxygenation), producing a product that cannot be used within the Calvin-Benson cycle. This process reduces efficiency of photosynthesis, potentially reducing photosynthetic output by 25% in C_3 plants. Photorespiration involves a complex network of enzyme reactions that exchange metabolites between chloroplasts, leaf peroxisomes and mitochondria.

The oxygenation reaction of RuBisCO is a wasteful process because 3-Phosphoglycerate is created at a reduced rate and higher metabolic cost compared with RuBP carboxylase activity. While photorespiratory carbon cycling results in the formation of G3P eventually, there is still a net loss of carbon (around 25% of carbon fixed by photosynthesis is re-released as CO_2) and nitrogen, as ammonia.

The ammonia must be detoxified at a substantial cost to the cell. Photorespiration also incurs a direct cost of 2ATP and one NAD(P)H. While it is common to refer to the entire process as photorespiration, technically the term refers only to the metabolic network which acts to rescue the products of the oxygenation reaction (phosphoglycolate).

Photorespiratory Reactions

Addition of molecular oxygen to ribulose-1,5-bisphosphate produces 3-phosphoglycerate (PGA) and 2-phosphoglycolate (2PG, or PG). PGA is the normal product of carboxylation, and productively enters the Calvin cycle. Phosphoglycolate, however, inhibits certain enzymes involved in photosynthetic carbon fixation (hence is often said to be an 'inhibitor of photosynthesis'). It is also relatively difficult to recycle: in higher plants it is salvaged by a series of reactions in the peroxisome, mitochondria, and again in the peroxisome where it is converted into glycerate. Glycerate reenters the chloroplast and by the same transporter that exports glycolate. A cost of 1 ATP is associated with conversion to 3-phosphoglycerate (PGA) (Phosphorylation), within the chloroplast, which is then free to re-enter the Calvin cycle.

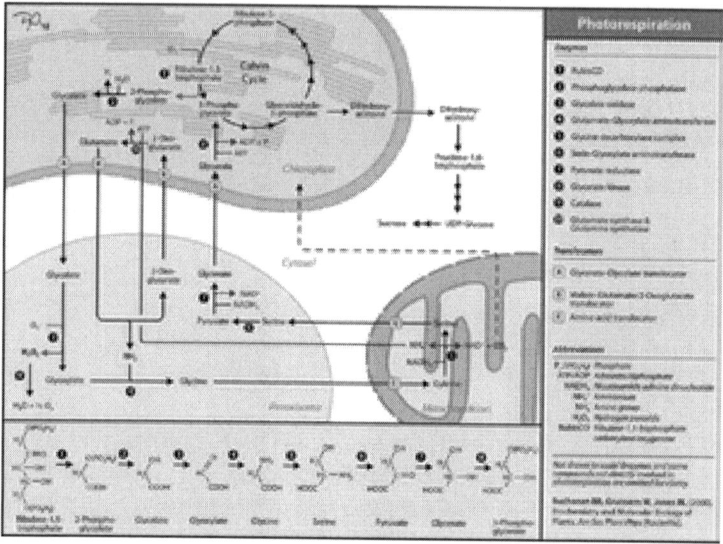

Figure: *Photorespiration*

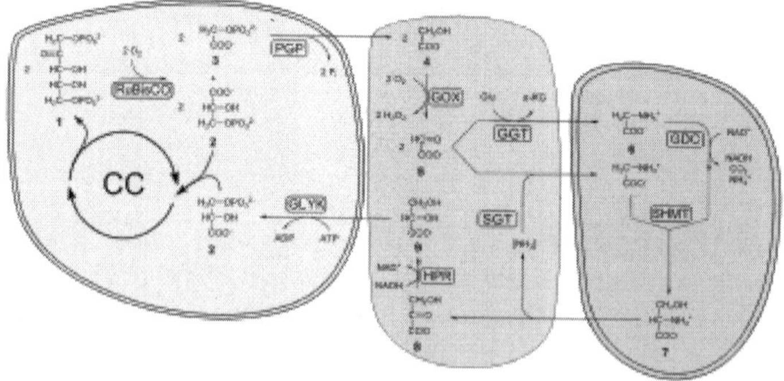

Figure: *Photorespiration*

There are several costs associated with this metabolic pathway; one being the production of hydrogen peroxide in the peroxisome (associated with the conversion of glycolate to glyoxylate). Hydrogen peroxide is a dangerously strong oxidant which must be immediately broken down into water and oxygen by the enzyme catalase. The conversion of 2x 2Carbon glycine to 1 C3 serine in the mitochondria by the enzyme glycine-decarboxylase is a key step, which releases CO_2, NH_3, and reduces NAD to NADH. Thus, 1 CO_2 molecule is produced for every 3 molecules of O_2 (two deriving from the activity of RuBisCO, the third from peroxisomal oxidations). The assimilation of NH_3 occurs via the GS-GOGAT cycle, at a cost of one ATP and one NADPH.

Cyanobacteria have three possible pathways through which they can metabolise 2-phosphoglycolate. They are unable to grow if all three pathways are knocked out, despite having a carbon concentrating mechanism that should dramatically reduce the rate of photorespiration.

Substrate Specificity of RuBisCO

The oxidative photosynthetic carbon cycle reaction is catalyzed by RuBP oxygenase activity:

RuBP + O_2 → Phosphoglycolate + 3-phosphoglycerate + 2H+

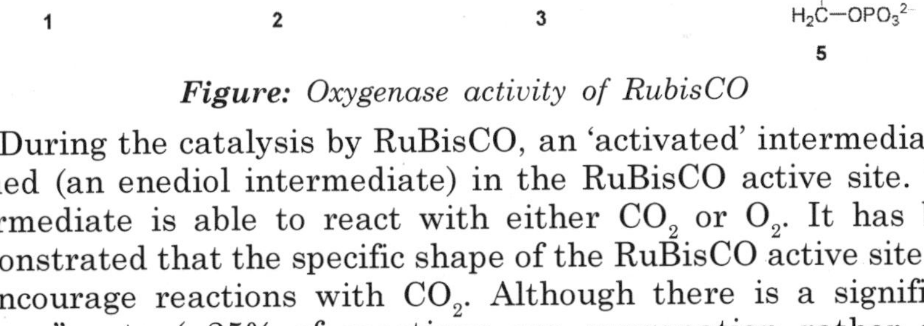

***Figure:** Oxygenase activity of RubisCO*

During the catalysis by RuBisCO, an 'activated' intermediate is formed (an enediol intermediate) in the RuBisCO active site. This intermediate is able to react with either CO_2 or O_2. It has been demonstrated that the specific shape of the RuBisCO active site acts to encourage reactions with CO_2. Although there is a significant "failure" rate (~25% of reactions are oxygenation rather than carboxylation), this represents significant favouring of CO_2, when the relative abundance of the two gasses is taken into account: in the current atmosphere, O_2 is approximately 500 times more abundant, and in solution O_2 is 25X more abundant than CO_2. The ability of RuBisCO to specify between the two gasses is known as its selectivity factor (or Srel), and it varies between species (although there is not much variation within the vascular plants).

A suggested explanation into Rubisco's inability to discriminate completely between CO_2 and O_2 is that it is an evolutionary relic: The early atmosphere in which primitive plants originated contained very little oxygen, the early evolution of RuBisCO was not influenced by its ability to discriminate between O_2 and carbon dioxide.

Conditions Which Increase Photorespiration

Photorespiration rates are increased by:

Altered Substrate Availability: Lowered CO2 or Increased O2

Factors which influence this include the atmospheric abundance of the two gases, the supply of the gases to the site of fixation (i.e. in land plants: whether the stomata are open or closed), the length of the liquid phase (how far these gases have to diffuse through water in order to reach the reaction site). For example, when the stomata are closed to prevent water loss during drought: this limits the CO_2 supply, while O_2 production within the leaf will continue. In algae (and plants which photosynthesise underwater) gases have to diffuse significant distances through water, which results in a decrease in the availability of CO_2 relative to O_2. It has been predicted that the increase in ambient CO_2 concentrations predicted over the next 100 years may reduce the rate of photorespiration in most plants by around 50%.

Increased Temperature

At higher temperatures RuBisCO is less able to discriminate between CO_2 and O_2. This is because the enediol intermediate is less stable. Increasing temperatures also reduce the solubility of CO_2, thus reducing the concentration of CO_2 relative to O_2 in the chloroplast.

High Light

Photorespiration may be used as a mechanism to dissipate excess energy at high irradiance levels.

Biological adaptation to minimize photorespiration

Certain species of plants or algae have mechanisms to reduce uptake of molecular oxygen by RuBisCO. These are commonly referred to as Carbon Concentrating Mechanisms (CCMs), as they increase the concentration of CO_2 so that Rubisco is less likely to produce glycolate through reaction with O_2.

Biochemical Carbon Concentrating Mechanisms

Biochemical CCMs concentrate carbon dioxide in one temporal or spacial region, through metabolite exchange. C4 and CAM photosynthesis both use the enzyme Phosphoenolpyruvate carboxylase (PEPC) to add CO_2 to a 3-Carbon sugar. PEPC is faster than Rubisco, and more selective for CO_2.

C4

C4 plants capture carbon dioxide in their mesophyll cells (using an enzyme called Phosphoenolpyruvate carboxylase which catalyzes the combination of carbon dioxide with a compound called Phosphoenolpyruvate (PEP)), forming oxaloacetate. This oxaloacetate is then converted to malate and is transported into the bundle sheath cells (site of carbon dioxide fixation by RuBisCO) where oxygen concentration is low to avoid photorespiration. Here, carbon dioxide is removed from the malate and combined with RuBP by Rubisco in the usual way, and the Calvin cycle proceeds as normal. The CO_2 concentrations in the Bundle Sheath are approximately 10-20 fold higher than the concentration in the mesophyll cells.

This ability to avoid photorespiration makes these plants more hardy than other plants in dry and hot environments, wherein stomata are closed and internal carbon dioxide levels are low. Under these conditions, photorespiration does occur in C4 plants, but at a much reduced level compared with C3 plants in the same conditions. C4 plants include sugar cane, corn (maize), and sorghum.

CAM (Crassulacean Acid Metabolism)

CAM plants, such as cacti and succulent plants, also use the enzyme PEP carboxylase to capture carbon dioxide, but only at night. Crassulacean acid metabolism allows plants to conduct most of their gas exchange in the cooler night-time air, sequestering carbon in 4 carbon sugars which can be released to the photosynthesizing cells during the day. This allows CAM plants to reduce water loss (transpiration) by maintaining closed stomata during the day. CAM plants usually display other water-saving characteristics, such as thick cuticles, stomata with small apertures, and typically lose around 1/3 of the amount of water per CO_2 fixed.

Algae

There have been some reports of algae operating a biochemical CCM: shuttling metabolites within single cells to concentrate CO2 in one area. This process is not fully understood and may not be biologically widespread.

Biophysical Carbon Concentrating Mechanisms

This type of carbon concentrating mechanism (CCM) relies on a contained compartment within the cell into which CO2 is shuttled, and where RuBisCO is highly expressed. In many species, biophysical CCMs are only induced under low carbon dioxide concentrations.

Biophysical CCMs are more evolutionarily ancient than biochemical CCMs. There is some debate as to when biophysical CCMs first evolved, but it is likely to have been during a period of low carbon dioxide, after the great oxidation event (2.4 billion years ago). Low CO_2 periods occurred around 750, 650, and 320-270 million years ago.

Eukaryotic Algae

In nearly all species of eukaryotic algae (*Chloromonas* being one notable exception), upon induction of the CCM, ~95% of RuBisCO is densely packed into a single subcellular compartment: the pyrenoid. Carbon dioxide is concentrated in this compartment using a combination of CO_2 pumps, bicarbonate pumps, and carbonic anhydrases. The pyrenoid is not a membrane bound compartment, but is found within the chloroplast, often surrounded by a starch sheath (which is not thought to serve a function in the CCM).

Hornworts

Certain species of Hornwort are the only land plants which are known to have a biophysical CCM involving concentration of carbon dioxide within pyrenoids in their chloroplasts.

Cyanobacteria

Cyanobacterial CCMs are similar in principle to those found in Eukaryotic algae and Hornworts, but the compartment into which carbon dioxide is concentrated has several structural differences. Instead of the pyrenoid, cyanobacteria contain Carboxysomes, which have a protein shell, and linker proteins packing RuBisCO inside with a very regular structure. Cyanobacterial CCMs are much better understood than those found in Eukaryotes, partly due to the ease of genetic manipulation of prokaryotes.

Photorespiration May Serve Some Purpose

Reducing photorespiration may not result in increased growth rates for plants. Some research has suggested, for example, that photorespiration may be necessary for the assimilation of nitrate from soil. Thus, a reduction in photorespiration by genetic engineering or because of increasing atmospheric carbon dioxide due to fossil fuel burning may not benefit plants as has been proposed. Several physiological processes may be responsible for linking photorespiration and nitrogen assimilation: one is that photorespiration increases availability of NADH, which is required for the conversion of nitrate to nitrite; another link is that certain nitrite transporters also transport bicarbonate, and elevated CO_2 has been shown to suppress nitrite transport into chloroplasts. Although photorespiration is greatly reduced in C4 species, it is still an essential pathway - mutants without functioning 2-phosphoglycolate metabolism cannot grow in normal conditions. One mutant was shown to rapidly accumulate glycolate.

Although the functions of photorespiration remain controversial, it is widely accepted that this pathway influences a wide range of processes from bioenergetics, photosystem II function, and carbon metabolism to nitrogen assimilation and respiration. The photorespiratory pathway is a major source of H_2O_2 in photosynthetic cells. Through H_2O_2 production and pyridine nucleotide interactions, photorespiration makes a key contribution to cellular redox homeostasis. In so doing, it influences multiple signalling pathways, in particular, those that govern plant hormonal responses controlling growth, environmental and defence responses, and programmed cell death.

Another theory postulates that it may function as a "safety valve", preventing the excess of reductive potential coming from an over-reduced NADPH-pool from reacting with oxygen and producing free radicals, as these can damage the metabolic functions of the cell by subsequent oxidation of membrane lipids, proteins or nucleotides.

Factors

Figure: *The leaf is the primary site of photosynthesis in plants.*

There are three main factors affecting photosynthesis and several corollary factors. The three main are:

- Light irradiance and wavelength
- Carbon dioxide concentration
- Temperature.

Light Intensity (Irradiance), Wavelength and Temperature

In the early 20th century, Frederick Blackman and Gabrielle Matthaei investigated the effects of light intensity (irradiance) and temperature on the rate of carbon assimilation.

- At constant temperature, the rate of carbon assimilation varies with irradiance, initially increasing as the irradiance increases. However, at higher irradiance, this relationship no longer holds and the rate of carbon assimilation reaches a plateau.
- At constant irradiance, the rate of carbon assimilation increases as the temperature is increased over a limited range. This effect is seen only at high irradiance levels. At low irradiance, increasing the temperature has little influence on the rate of carbon assimilation.

These two experiments illustrate vital points: First, from research it is known that, in general, photochemical reactions are not affected by temperature. However, these experiments clearly show that temperature affects the rate of carbon assimilation, so there must be two sets of reactions in the full process of carbon assimilation. These are, of course, the light-dependent 'photochemical' stage and the light-independent, temperature-dependent stage. Second, Blackman's experiments illustrate the concept of limiting factors. Another limiting factor is the wavelength of light. Cyanobacteria, which reside several metres underwater, cannot receive the correct wavelengths required to cause photoinduced charge separation in conventional photosynthetic pigments. To combat this problem, a series of proteins with different pigments surround the reaction centre. This unit is called a phycobilisome.

Carbon Dioxide Levels and Photorespiration

As carbon dioxide concentrations rise, the rate at which sugars are made by the light-independent reactions increases until limited by other factors. RuBisCO, the enzyme that captures carbon dioxide in the light-independent reactions, has a binding affinity for both carbon dioxide and oxygen. When the concentration of carbon dioxide is high, RuBisCO will fix carbon dioxide. However, if the carbon dioxide concentration is low, RuBisCO will bind oxygen instead of carbon dioxide. This process, called photorespiration, uses energy, but does not produce sugars.

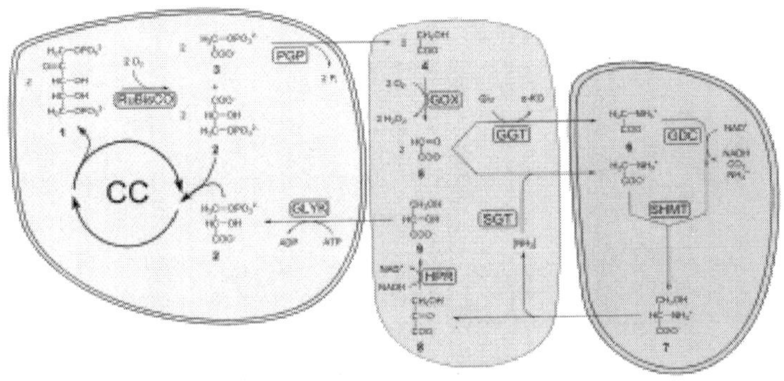

Figure: *Photorespiration*

RuBisCO oxygenase activity is disadvantageous to plants for several reasons:

1. One product of oxygenase activity is phosphoglycolate (2 carbon) instead of 3-phosphoglycerate (3 carbon). Phosphoglycolate cannot be metabolized by the Calvin-Benson cycle and represents carbon lost from the cycle. A high oxygenase activity, therefore, drains the sugars that are required to recycle ribulose 5-bisphosphate and for the continuation of the Calvin-Benson cycle.
2. Phosphoglycolate is quickly metabolized to glycolate that is toxic to a plant at a high concentration; it inhibits photosynthesis.
3. Salvaging glycolate is an energetically expensive process that uses the glycolate pathway, and only 75% of the carbon is returned to the Calvin-Benson cycle as 3-phosphoglycerate. The reactions also produce ammonia (NH_3), which is able to diffuse out of the plant, leading to a loss of nitrogen.

A highly simplified summary is: 2 glycolate + ATP → 3-phosphoglycerate + carbon dioxide + ADP + NH_3

The salvaging pathway for the products of RuBisCO oxygenase activity is more commonly known as photorespiration, since it is characterized by light-dependent oxygen consumption and the release of carbon dioxide.

7

Keeping Cool

Plants have many important functions, like making leaves, making flowers and seeds, growing, storing starches in the roots etc, but we humans are usually unaware of the vital function of transpiration.

It is estimated that 98% of a plants energy is used in the work of transpiration. How does this process work? and why is it so essential to a plant?

Water moves from the soil into plant roots, up through the sapwood into the leaves. The water, warmed by the sun, turns into vapor (evaporates), and passes out through thousands of tiny pores (stomata) mostly on the underside of the leaf surface. This is transpiration. It has two main functions: cooling the plant and pumping water and minerals to the leaves for photosynthesis.

Plants need to cool themselves for several reasons. When temperatures are too high, energy systems (metabolic functions) slow, and growth and flowering slows or stops. In extreme heat, plants are severely stressed and can die. Sometimes heat will cause bubbles to form that block the flow of water, leading to dehydration. Transpiration is an evaporative cooling system that brings down the temperature of plants, but since it leads to water loss, it must be accurately regulated. The ingenious system that regulates this function consists of a guard cell on each side of the tiny pores (stomata). When water moves into the guard cells, they swell and arch open; as water moves out, the guard cells relax and close. The guard cells are sensitive to light intensity, temperature, wind, relative humidity and carbon dioxide concentrations inside the leaf.

The stomata (pores) must open to take in carbon dioxide for photosynthesis (especially important on mornings of sunny days). And the more they are open, the more plants transpire and lose water. So watering plants early in the morning will support plant energy, especially on hot summer days.

What causes water to rise up through a plant (sometimes 100 feet or more) against the force of gravity? This is achieved by the pumping action of transpiration, which is another ingenious system. It depends on the tiny (capillary) xylem water channels, the extremely strong cohesive (bonding) properties of water molecules, and a sucking force created when water at the top of the channels evaporates. Like sucking on a soda straw, transpiration causes a negative pressure which lifts the column of water to the leaf surface.

As plants transpire, the soil becomes dryer. Then in summer, if the soil becomes very dry, plants cannot transpire enough to keep cool. In desperation plants may start losing leaves or completely defoliate as a way to keep transpiration from dehydrating the plant. Here in the West, in order to help our plants keep cool and keep them photosynthesizing, we can give them extra water in the hottest, driest months.

We can plant wind breaks and trees and shrubs that will offer a little shade, and we can mulch to keep soil moisture from evaporating. Evergreens, especially broadleaf evergreens, are very vulnerable to the West's intense winter sun. Because they have leaves, they continue to transpire in the winter, and can dehydrate. This is especially true for any evergreen that is planted in the fall, but critical for fall-planted ball-and-burlap evergreen trees. So water evergreens once or twice a month in winter.

Transpiration is an elegant, sustainable natural design that performs its vital functions without electricity, without fossil fuels and without moving parts. It does not pollute or use excessive resources. It actually adds moisture to the atmosphere and contributes to rainfall.

Other interesting facts about Transpiration:

1. A leaf transpires about 90% of the water evaporated from a water surface of the same area—even though the combined area of stomatal pores is only 1-2% of the total leaf area.
2. Transpiration rates are highest in leaves that are stiff with turgor (water pressure). When leaves wilt, they offer less surface area to sun exposure, and thus will transpire less, saving water. Watch a tough, drought-tolerant plant like lilac when temperatures are high for a demonstration of this water-saving strategy.

3. Succulents save water by opening the stomata pores at night to reduce transpiration and to take in carbon dioxide which is stored in their leaves until the next day when they can photosynthesize.
4. Cacti, since they don't have leaves, only have a few stomata in their green stems and so transpire very little.
5. Many xeric plants have small leaves, silvery reflective leaves, hairy leaves and/or produce essential oils which are all strategies to reduce transpiration by reducing evaporation.
6. In the summer, a large maple tree can transpire 50-60 gallons of water per hour into the atmosphere. This adds to the humidity which in the West helps us to feel more comfortable, is less drying for our skin and reduces bronchial problems in our lungs. It also helps to cool our environment. Where there are large areas of trees, the combined effect of their transpiration can create a super-saturated condition in the clouds, which can result in rain or snow. The Snows of Kilimanjaro have disappeared primarily because deforestation has reduced local transpiration, resulting in less (or no) snowfall.
7. Anti-transpirant products, like Wilt-Pruf, can reduce transpiration by covering the stomata pores, but that also limits photosynthesis. They can be valuable sprayed on evergreens to get them through their first winter in your landscape.

Keep It Cool: What Desert Plants can Teach Us about Climate Change

Climate change isn't new. The earth's organisms have faced the "cope, adapt, or die" paradigm presented by changing climatic conditions since the inception of life on earth more than 3 billion years ago.

Some species already live in extreme conditions (a topic I covered in my last blog post) – at the earth's freezing poles in the dead of winter, at the apex of tropical mountain peaks that receive tens of metres of rainfall a year, and deep in the dry deserts that comprise a third of the earth's land area. The plants and animals able to survive these extreme conditions may have something to teach us about climate change adaptation—especially when the organisms are incapable of crawling up to a cooler elevation or slithering into a deeper riverine pool to escape a heat wave.

As a general rule, plants spend their lives rooted to a single spot on the earth. So they need to be able to withstand 365 days per year of whatever nature can dish out in terms of weather. Plants from hot deserts—such as the prickly cholla cactus—are no exception, and they have evolved a broad range of impressive strategies for coping with the hottest, driest climate in North America, where a years' worth of rainfall may come in a single extreme event.

These desert denizens provide us with valuable insight into biological and physical adaptations that allow for survival on a hotter, drier planet that is subject to extreme events (as climate change experts are currently forecasting).

In addition, deserts plants hold fascinating keys to longevity: the California deserts are home to both the world's oldest tree—a Great Basin bristlecone pine (*Pinus longaeva*) between 5,062 and 5,063 years old—and "King Clone", the oldest clonal colony of creosote bush (*Larrea tridentata*), which is estimated to be 11,700 years of age.

From the perspective of climate change and adaptation, the ancient creosote clone is particularly intriguing, as the climate under which it originally germinated (at the tail end of the last ice age) was different from that in which it lives today.

What Strategies Allow Desert Plants to Survive Harsh Conditions?

Well, plants protect themselves from intense heat by producing smaller leaves (spines in cactus), by using water-saving methods of photosynthesis (such as Crassulacean acid metabolism), by growing protective hairs to deflect sunlight, or by producing thin leaves that cool down easily in a breeze or waxy leaves that prevent water loss.

They can also capture moisture by having short roots that expand when it rains, or extremely long, fast-growing roots that can quickly tap into groundwater.

Cacti contain flexible structures that allow their stems to expand and store extra water to use when it isn't raining. Finally, plants have evolved the ability to delay germination and growth to coincide with water availability and mild temperatures, thereby avoiding the exposure of tender young seedlings to the harshest conditions.

You may be thinking at this point: these adaptations are great, but why should I care?

Beyond their intrinsic beauty and their support of other desert species, plants have tangible benefits to offer society. Using biomimicry, scientists and engineers have begun to copy the strategies of animals and plants in nature in order to solve human problems. Desert plants may serve as particularly helpful guides as we attempt to adapt to a hotter and drier planet. We'll have the best chance to learn from them if we protect the habitats where they are found.

How to Help Your Crops Endure Hot Weather

Hot weather is tougher on plants than it is on people. It's easy to understand why, when you consider that our bodies contain about 60 percent water and most plants are 85 to 90 percent water. So when temperatures rise, plants get even thirstier and sweatier than we do.

As with people, some plants tolerate heat better than others. Knowing which plants like it hot and which would prefer air conditioning, you can help your vegetables and flowers survive — and even thrive in — hot weather.

Vegetables: Tomatoes, peppers, eggplant, corn, melons and squash actually need at least a month of 80 to 90 degree weather to develop a flavourful and abundant crop. As long as they don't run out of water, these sun worshippers are well equipped to survive the heat. On hot days they conserve energy and moisture by slowing down. While resting, their foliage may appear to be wilting from lack of water, but as evening approaches they'll perk up again. Heat loving plants are thirsty– the average tomato plant needs more than 30 gallons of water in a season. Using a combination of mulch and drip or soaker hose will ensure these plants stay healthy and well hydrated.

In the vegetable garden, it's the cool weather crops — lettuce, spinach, arugula, broccoli, kale, cauliflower, peas, cilantro — that suffer in hot weather. Even with an abundant and consistent supply of water, when temperatures rise over 80 degrees, these plants tend to stop growing, go to seed, or just give up the ghost.

Cool-weather crops should be grown on either side of midsummer heat. The plants will be much healthier and they'll taste better, too. Look for varieties that are well matched to spring or fall production. Plan to sow salad greens every couple weeks to maintain a steady supply of high-quality leaves. To learn more about succession planting, read Double Your Harvest with a Second Planting and The Vegetable Garden in Autumn.

In hot weather, heat-sensitive crops want protection from both heat and sun. Shade netting helps these plants in several ways. Most importantly, it keeps soil and air temperatures as much as 10 degrees cooler (lettuce seeds germinate poorly in soil temperature higher than 70 degrees). Shade netting also protects tender foliage from being scorched by intense sunlight and reduces moisture loss. For best results, suspend the shade netting several inches above your crops, letting it rest on wire or fiberglass hoops, or a wooden frame.

Flowers

Perennials that bloom in mid and late summer, such as asters, echinacea, rudbeckia, sedum and daylilies, are usually unfazed by heat. They may need extra water when the thermometre hits 90 degrees, but the heat itself doesn't bother them much.

Cool-weather perennials that flower in spring and early summer often have a hard time with heat. These get their blooming work out of the way early so they can kick back when temperatures gets hot. In fact, some cool-weather perennials, such as Oriental poppies and bleeding heart, avoid summer altogether; they die right to the ground after blooming and don't reappear until next spring. But for the most part, these plants just struggle along, looking gangly and unkempt. You can help them out by removing spent blossoms and stems as soon as they've finished flowering. Trim back foliage that's not fresh and healthy. If you haven't done so already, mulch around the plants to keep the soil cool and water as needed during hot weather to keep the soil from drying out. A piece of shade netting, laid right on the foliage, will provide instant relief for plants that have been damaged, transplanted or are otherwise ailing.

Annuals span the full range from heat lovers (gazania and portulaca) to heat haters (pansies and sweet peas), so it's impossible to generalize about their care. With so many annuals to choose from, the best strategy is to match plant to place. If the snapdragons and petunias along your front walk are struggling in the heat, next year consider plants that have a higher tolerance for heat and drought, such as zinnias or gaillardia.

For annuals grown in pots, hot weather can be deadly. Ignore them for a couple days and they may never revive. When you water, make sure that the entire root ball gets moistened. Daily watering quickly leaches nutrients from the soil, so either use a slow-release fertilizer or apply a water-soluble liquid fertilizer every other week throughout the season. Nip off spent blooms and trim back rangy stems to encourage reblooming.

Many annuals and perennials now have a USDA heat zone rating as well as a hardiness rating. When a plant is not cold hardy, it will simply die, but the signs of heat stress are usually more subtle. Plants that can't tolerate high heat may stop blooming, the leaves may turn pale, and the plants may become more susceptible to pests.

8

Cleaning Surfaces

Lotus Effect

The lotus effect refers to the very high water repellence (superhydrophobicity) resulting in self-cleaning properties, as exhibited by the leaves of the lotus flower (*Nelumbo*). Dirt particles are picked up by water droplets due to a complex micro- and nanoscopic architecture on the surface, which minimizes the droplet's adhesion to said surface. Superhydrophobic and self-cleaning properties can also easily be demonstrated in many other plants, for example *Tropaeolum* (nasturtium), *Opuntia* (prickly pear), *Alchemilla*, cane, and on the wings of certain insects.

The phenomenon of superhydrophobicity was first studied by Dettre and Johnson in 1964 using rough hydrophobic surfaces. Their work developed a theoretical model based on experiments with glass beads coated with paraffin or PTFE telomer. The self-cleaning property of superhydrophobic micro-nanostructured surfaces was studied by Barthlott and Ehler in 1977, who described such self-cleaning and superhydrophobic properties for the first time as the "lotus effect"; perfluoroalkyl and perfluoropolyether superhydrophobic materials were developed by Brown in 1986 for handling chemical and biological fluids. Other biotechnical applications have emerged since the 1990s.

Functional Principle

Due to their high surface tension, water droplets tend to minimize their surface by trying to achieve a spherical shape. On contact with a surface, adhesion forces result in wetting of the surface. Either complete or incomplete wetting may occur depending on the structure of the surface and the fluid tension of the droplet.

The cause of self-cleaning properties is the hydrophobic water-repellent double structure of the surface. This enables the contact area and the adhesion force between surface and droplet to be significantly reduced resulting in a self-cleaning process.

This hierarchical double structure is formed out of a characteristic epidermis (its outermost layer called the cuticle) and the covering waxes. The epidermis of the lotus plant possesses papillae with 10 to 20 μm in height and 10 to 15 μm in width on which the so-called epicuticular waxes are imposed. These superimposed waxes are hydrophobic and form the second layer of the double structure.

The hydrophobicity of a surface can be measured by its contact angle. The higher the contact angle the higher the hydrophobicity of a surface. Surfaces with a contact angle $< 90°$ are referred to as hydrophilic and those with an angle $>90°$ as hydrophobic.

Some plants show contact angles up to 160° and are called super-hydrophobic meaning that only 2–3% of a drop's surface is in contact. Plants with a double structured surface like the lotus can reach a contact angle of 170° whereas a droplet's actual contact area is only 0.6%. All this leads to a self-cleaning effect. Dirt particles with an extremely reduced contact area are picked up by water droplets and are thus easily cleaned off the surface.

If a water droplet rolls across such a contaminated surface the adhesion between the dirt particle, irrespective of its chemistry, and the droplet is higher than between the particle and the surface. As this self-cleaning effect is based on the high surface tension of water it does not work with organic solvents. Therefore, the hydrophobicity of a surface is no protection against graffiti.

This effect is of a great importance for plants as a protection against pathogens like fungi or algae growth, and also for animals like butterflies, dragonflies and other insects not able to cleanse all their body parts. Another positive effect of self-cleaning is the prevention of contamination of the area of a plant surface exposed to light resulting in a reduced photosynthesis.

Technical Application

When it was discovered that self-cleaning qualities come from the physical-chemical properties of superhydrophobic surfaces at the microscopic to nanoscopic scale, and does not result from any of the specific chemical properties of the surface of the leaf, that discovery opened up the possibility of using this effect in our own manmade surfaces, by mimicking nature in a general way rather than a specific one. Some nanotechnologists have developed treatments, coatings, paints, roof tiles, fabrics and other surfaces that can stay dry and clean themselves by replicating in a technical manner the self-cleaning properties of plants, such as the lotus plant.

This can usually be achieved using special fluorochemical or silicone treatments on structured surfaces or with compositions containing micro-scale particulates. Super-hydrophobic coatings comprising Teflon microparticles have been used on medical diagnostic slides for over 30 years.

It is possible to achieve such effects by using combinations of polyethylene glycol with glucose and sucrose (or any insoluble particulate) in conjunction with a hydrophobic substance.

Further applications have been marketed, such as self-cleaning glasses installed in the sensors of traffic control units on German autobahns developed by a cooperation partner (Ferro GmbH). Evonik AG has developed a spray for generating self-cleaning films on various substrata.

Superhydrophobic coatings applied to microwave antennas can significantly reduce rain fade and the buildup of ice and snow. "Easy to clean" products in ads are often mistaken in the name of the self-cleaning properties of hydrophobic or superhydrophobic surfaces. Patterned superhydrophobic surfaces also show promise for "lab-on-a-chip" microfluidic devices and can greatly improve surface-based bioanalysis.

Superhydrophobic or hydrophobic properties have been used in dew harvesting, or the funneling of water to a basin for use in irrigation. The Groasis Waterboxx has a lid with a microscopic pyramidal structure, based on the superhydrophobic properties that funnel condensation and rainwater into a basin for release to a growing plant's roots.

Research History

Although the self-cleaning phenomenon of the lotus was possibly known in Asia long before (reference to the lotus effect is found in the Bhagavad Gita) its mechanism was explained only in the early 1970s after the introduction of the scanning electron microscope. Studies were performed with leaves of *Tropaeolum* and lotus (*Nelumbo*). The Lotus Effect is a registered trademark of Sto AG (US Registration Nos. 3561156 and 2613850).

How Nature Cleans

Learning How to Clean From a Leaf

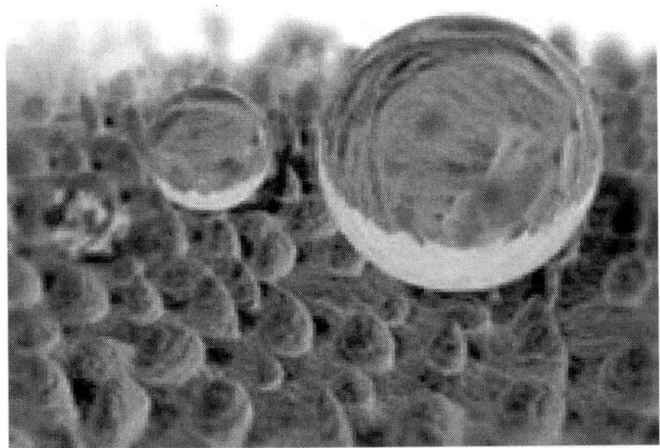

Have you ever wondered how things in nature generally look so *clean*? You don't see janitors out there in the woods, afterall, dusting off the trees. People use toxic detergents and costly cleaning treatments, but Nature employs a cleaning strategy as environmentally benign and energy efficient as it is strikingly ingenious. Imagine waking up, stepping outside, shaking your body a little bit, and heading off to your daily routine as clean as if you'd taken a shower. Because that's how Nature cleans; it takes what might be called a *gravity shower*.

The leaves of many plants, large-winged insects, most water birds, and other organisms capitalize on basic physical characteristics in the way surfaces of materials interact, achieving cleanliness effortlessly and without detergents.

How It Works

Water has two major forces acting on it at the same time: its attraction to itself (which causes it to ball up) and adhesive forces (which cause it to stick to surfaces, pulling it down). Adhesive forces on water tend to be maximized on smooth surfaces, because the liquid-to-solid contact area is large.

But because water and air are much less naturally adhesive than water and solids, roughened surfaces of certain microstructure tend to reduce adhesive force on water droplets, as trapped air in the interstitial spaces of the roughened surface result in a reduced liquid-to-solid contact area. This allows water's self-attraction to be expressed more fully, leading it to form a sphere.

At the same time, due again to natural adhesion between water and solids, dirt particles on a leaf's surface stick to the water, like a rolled snowball that picks up leaves from your lawn. Since a ball rolls more easily than a flattened bump, the role of gravity now becomes significant: the slightest angle in the surface of the leaf (e.g., caused by a passing breeze) causes balls of water to roll off the leaf surface, carrying away the attached dirt particles - without the leaf having to expend any energy or use any harmful chemical detergents.

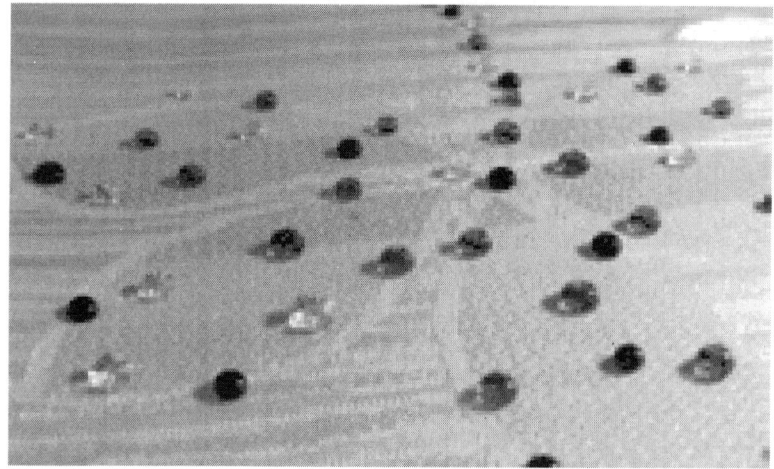

GreenShield, a fabric finish inspired by the Lotus Effect from G3 Technology Innovations, coats textile fibers with liquid-repelling nanoparticles, reducing the need to use environmentally dangerous fluorocarbons.

Why It Matters

Nature's way of cleaning has major implications for the world we live in. Considering how damaging to the environment and our own health our cleaning practices are — as well as how financially costly - we stand to gain a great deal from learning from the innovative cleaning strategies employed by Nature. Research and development of the "Lotus Effect" is actively underway.

Several products exist already which use the Lotus Effect to improve on conventional products. GreenShield™, a product from G3 Technology Innovations, is a fabric finish that uses the principle of the Lotus Effect to create water and stain repellency on textiles, and results in a 10-fold decrease in the use of environmentally harmful fluorocarbons, the conventional means of achieving repellency. Other products inspired by the Lotus Effect include Lotusan paint and Signapur glass finish.

Rice Leaves and Butterfly Wings Provide Insight into Nature's Best Self-Cleaning Surfaces

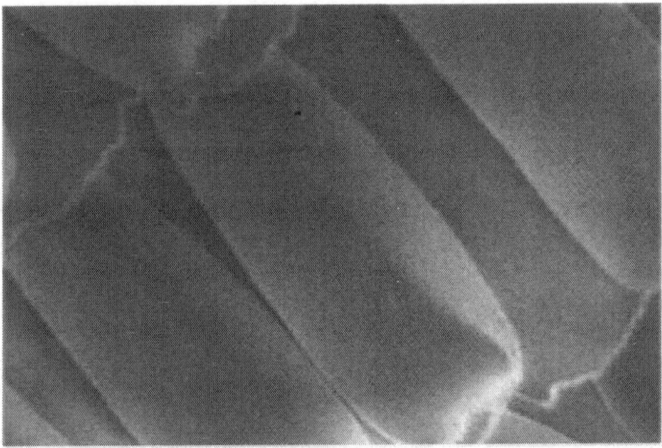

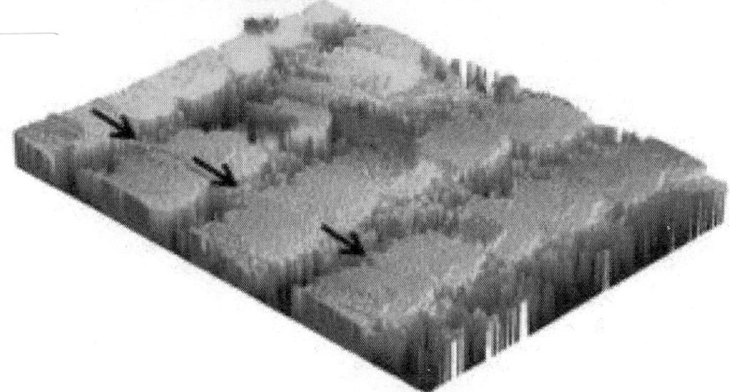

Figure: *Enlarge*

Butterfly wings are characterized by shingle-like scales (above), which keep water from pooling, and microgrooves (below), which repel water and reduce surface adhesion. These features help keep the surface free of debris and contaminants. ...more

With 3.5 billion years of research and development under her belt, Mother Nature could be considered the world's most experienced biological engineer. Sure, her methods may appear haphazard at times, but her track record of developing organisms that are exquisitely adapted to the tasks required of them is nothing short of amazing.

One task she's been particularly devoted to is finding ways to keep her creations clean of debris and contaminants. It's therefore not surprising that today's engineers look to nature for inspiration when it comes to dealing with "biofouling," or the unwanted build-up of biological material, which plagues a wide range of industries.

A recent study conducted by researchers at Ohio State University has found that rice leaves and butterfly wings make use of some unique surface characteristics that promote self-cleaning. The researchers believe that incorporating some of these features into man-made products might be key to tackling problems associated with biofouling.

"Living nature is full of engineering marvels, from the micro to the macro scale, that have inspired mankind for centuries," says Bharat Bhushan, senior author of the study and director of the Nanoprobe Laboratory for Bio- and Nanotechnology and Biomimetics at Ohio State University.

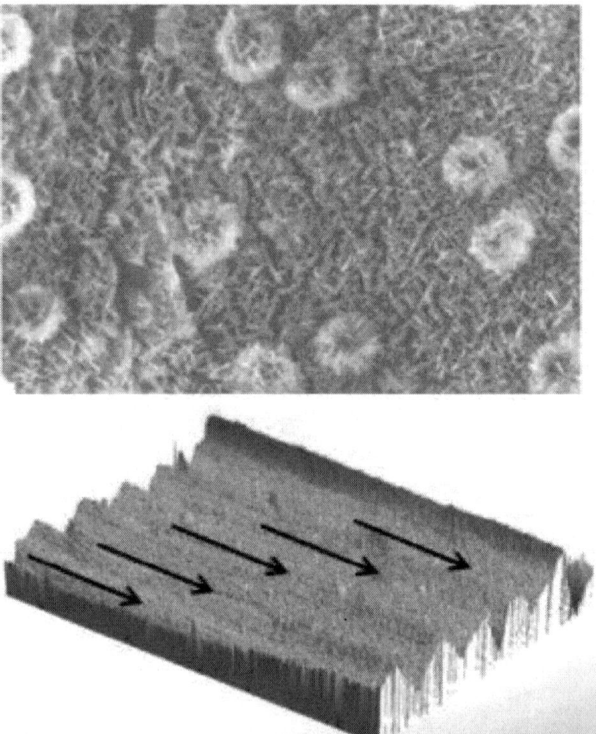

Figure: *Enlarge*

Consider, for example, that while a ship gets bogged down by barnacles as it crosses the ocean, a shark swimming in the same ocean remains clean as a whistle. One reason for this is that shark skin is composed of a special type of scale covered by riblets that reduce friction as the shark travels through the water. Reduced friction means that water flows more rapidly across the surface, making it difficult for microscopic hitchhikers to grab hold. This phenomenon is called the shark-skin effect. The lotus leaf, on the other hand, maintains its squeaky cleanness with a waxy surface structure that repels water, a property called superhydrophobicity. Combined with low adhesion, this is known as the lotus effect.

The September 2012 study, published in the journal *Soft Matter* by Bhushan and engineering graduate student Greg Bixler, shows that rice leaves and butterfly wings combine the low drag of shark skin with the superhydrophobicity of the lotus leaf, putting these surfaces at the top of the list of nature-made self-cleaners.

Rice leaves are characterized by waxy bumps (above), which repel water and prevent contaminants from adhering to the surface, as well as grooves (below) that keep water from pooling on the surface. Incorporating these features into man-made ...more

The idea to look at rice leaves and butterfly wings came to the investigators from observing these structures in their natural habitats.

"We noticed that water droplets on rice leaves and butterfly wings roll off effortlessly, and that each remains clean in their respective environment," says Bhushan.

This observation lead the investigators to suspect that, like shark skin and lotus leaves, rice leaves and butterfly wings have special properties that make them particularly resistant to fouling.

Before they could get started, they had to address the fact that both of these structures are incredibly delicate, making them difficult to work with in an experimental setting. For this reason, they began by creating replicas of both surfaces. Silicone was poured over actual leaves and wings, creating a "negative" mold they then used to create a urethane replica better suited to the rigorous tests the investigators had in mind. Some replicas also received a silica coating to replicate the superhydrophobic properties of the natural structures.

They then subjected the replicas to experiments designed to determine how efficiently they moved through air (drag), how well they got rid of contaminants (self-cleaning), how tightly contaminants stuck to the surface (adhesion), and how well they retained or repelled water (wettability). Like shark skin, rice leaves and butterfly wings exhibited low drag and self-cleaning properties.

But both of these samples were special in an important way: They exhibited lotus-like properties including superhydrophobicity and low adhesion. This effect was magnified in coated samples, which outperformed uncoated samples in every test.

Bixler attributes these findings to the unique structure of each surface.

"Both rice leaves and butterfly wings contain micro- and nano-sized features that repel and direct water in one direction," says Bixler. "This is accomplished with a combination of grooves and bumps that are invisible to the naked eye."

By showing that rice leaves and butterfly wings combine anti-fouling properties of some of nature's best self-cleaners, Bhushan and Bixler have identified new surfaces that can be used as engineering inspiration for a wide range of industries plagued by biofouling. Preventing the build-up of biological matter on a ship's hull, for example, could increase the efficiency of the ship's movement, ultimately leading to more efficient fuel use. Also, reducing the accumulation of bacteria and other microbes in medical tubing could greatly reduce a patient's risk of infection.

"We are investigating methods to fabricate rice leaf and butterfly wing-inspired films for applications requiring low drag, self-cleaning and anti-fouling," say Bhushan. The investigators hope that the use of such films in various industries, including health care, shipping and advanced manufacturing, will reduce costs and improve quality.

Lotus Leaves, Hydrophobic and Omniphobic Surfaces

Surfaces can be hydrophobic. Like Lotus leaves. They can be superhydrophobic. Like soot coated discs that bounce off milk drops. Now they can be omniphobic. Recent research demonstrate surfaces that can repel both water and oil.

When water droplet touches a surface, smaller its contact angle with the surface, better it would stick to the surface. Conversely, higher the contact angle, the droplet retains its shape and rolls down the surface (when it is inclined) without much friction. Like what happens to rain-drops on a lotus leaf.

Liquids wet surfaces when their surface tension is low, resulting in smaller contact angles for the liquid droplet with the surface. Higher the surface tension, better their ability to roll up into near spherical shapes, reducing wetting. Water's very high surface tension, 72 milliNewtons per metre (mN/m) at room temperature forms near-spherical drops when placed on a superhydrophobic surface with contact angle above 150°. Oils such as pentane have a low surface tension (15mN/m) hence low contact angle - so they wet the surface under gravity as a flat pool. Surfaces can be made to repel liquids by controlling their interface contact angle.

There are theoretical models available for how water contact with surfaces. In one it can be modelled as water touching only the irregular peaks of rough protrusions of the surface while in another model the water can stick to both the peaks and valleys of the roughness. Both configuration can form stable equilibrium but the former allows easy roll off of the droplet because of the air trapped in the gaps between the water droplet and the rough surface valleys.

Lotus leaves have a curious surface. They contain both micro-scale protrusions interspersed with nano-scale protrusions. When it sits on such surfaces, water is supported by the micro structure with the 'gaps' between the water droplet and the surface filled by air. This results in easy rolling off of the droplet.

Sometime back in a cute experiment, researchers studied the effect of the nano hair-like structure on the hydrophobic ability of lotus leaf. They measured the respective contact angles for untreated lotus leaf and lotus leaf annealed to shave off their nano-hair. Annealing is a slow process of heating and subsequent cooling of materials to change their micro-structure or stress relieving - in this case, removal of lotus leaf nano-hair - without changing much the chemical composition of the surface. I never suspected such a delicate nano-scale shaving process is possible with annealing.

Lotus leaf surface have high contact angles with water - as high as 140 to 170 degrees from the horizontal plane. Compared to this human skin, which is also slightly hydrophobic, has a contact angle of about 90 and wax coated smooth material surface about 75. In the above experiment, researchers showed that compared to the 170 degree contact angle of the untreated lotus leaf, the lotus leaf minus nano-hair structure offered only 126 degree contact angle, reducing the non-stick (and thereby, the self cleaning) ability of the lotus leaf.

That is in 2006. Now MIT researchers Robert Cohen and Gareth McKinley have come up with surfaces that are omniphobic that repel both water and oil (although 'omniphobic' means fear of anything). To illustrate the potential, they created surfaces that are impervious both to water and to alkanes - and modified duck feathers to make them oil-resistant.

According to the paper, due to its surface texture, the material allows water and oils to retain their spherical shape with their surface tension. Hence they both roll off easily. The surface testure is made up of 300-nanometre-tall "toadstools" with broad silicon dioxide caps and narrow silicon stems. The shape of the omniphobic toadstools makes it possible for even oils with that weak surface tension to hold as a droplet together. This allows liquids like pentane (oil) to form a sphere without collapsing. Similar to the lotus leaf effect discussed above, the oil droplet meniscus forms a large enough contact angle with the nano-toadstool textured surface, with the meniscus resting on a layer of air beneath the toadstools' peaks (caps).

Although self-cleaning solar panels are thought of as applications for these omniphobic materials, commercial use of such textures are expected. As such, these surface textures when coated on regular solids, get destroyed.

9

Staying Unfrozen

Pine trees, hemlocks and rhododendrons really stand out during the winter months because they stay green all year long. But how are they able to keep their leaves when most of the other plants in the forest have lost theirs? Evergreen shrubs and trees in northern climates face many challenges during winter. Cold temperatures are actually the least critical problem. Many plants can carry out photosynthesis with temperatures at or a few degrees below freezing.

The biggest problem for plants in winter is desiccation or drying out. When the ground freezes, plant roots are unable to draw liquid water from the soil. When the plant becomes dehydrated, the plant cell membranes lose their shape and rupture, killing portions of the plant. Cell tissues can also die if ice crystals form inside cell membranes.

Evergreen plants in West Virginia have many different ways to survive the cold and dry winters. Many evergreens have tough, leathery leaves or needles that are resistant to drying out. Some plants can actively concentrate chemicals that work like antifreeze in their cell fluids. For other plants, simply being buried in a snowbank can help reduce water loss. Snow provides moisture as well as shelter from drying winds. However, being buried under a thick layer of ice for too long may kill a plant because toxic concentrations of compounds like carbon dioxide and ethanol build up in the closed air space beneath the ice.

—John Beckman

Do Insects Hibernate?

Insects survive the winter by using a variety of methods to protect their cells from freezing and being killed. One common mechanism is the production of large amounts of glycerol, a natural antifreeze, in an insect's blood. As much as a quarter of the body weight of a "winter-hardy" insect may be made of glycerol which has the effect of greatly reducing the freezing point.

Other insects have means to partially control the freezing so that it occurs in the spaces outside the cells, protecting the cell walls. And almost all insects have behaviours that cause them to seek out protective sites for wintering, such as under sheltering debris or in similar insulated areas where temperature extremes are moderated.

Also, almost all insects that overwinter go into diapause, a state of arrested development and growth. This condition can only be "broken" if certain environmental conditions are met, such as a period of chilling has passed or a critical day length is perceived. Diapause has the protective effect of preventing many insects from prematurely emerging and losing winter hardiness during a warm spell.

Source: Colorado State University Cooperative Extension Service

Why Don't Fish Freeze?

Although humans would never be able to withstand a long period of time in cold water (remember the Titanic?), many freshwater fish survive because of special adaptations that help with internal thermal regulation. Fish, in contrast to warm-blooded mammals and birds which maintain a constant internal body temperature, are cold-blooded animals. Their internal body temperature equals that of the environment around them.

Freshwater fish avoid freezing because they are hyperosmotic, meaning they have a higher internal concentration of salts than that of their surrounding freshwater environment. This is important because it takes heavier or more concentrated liquids a longer time to freeze than less concentrated, just as it takes saltwater longer to freeze than freshwater.

To avoid getting too cold, fish use either behavioural habits to move from one water mass or area to another in search of warmer water, or physiological means to keep at the proper temperature. In the wintertime you may have experienced fish stacked up in one spring-fed pool. Spring water is approximately 50 degrees and provides a thermal refuge from colder water. Of course, all these methods won't help unless part of the fish's environment remains unfrozen.

Vernalization

Vernalization (from Latin: vernus, *of the spring*) is the acquisition of a plant's ability to flower in the spring by exposure to the prolonged cold of winter, or by an artificial equivalent. After vernalization, plants have acquired the ability to flower, but they may require additional seasonal cues or weeks of growth before they will actually flower.

Many plants grown in temperate climates require vernalization and must experience a period of low winter temperature to initiate or accelerate the flowering process. This ensures that reproductive development and seed production occurs in spring and summer, rather than in autumn. The needed cold is often expressed in chill hours. Typical vernalization temperatures are between 5 and 10 degrees Celsius (40 and 50 degrees Fahrenheit).

For many perennial plants, such as fruit tree species, a period of cold is needed first to induce dormancy and then later, after the requisite period of time, re-emerge from that dormancy prior to flowering. Many monocarpic annuals and biennials, including some ecotypes of *Arabidopsis thaliana* and winter cereals such as wheat, must go through a prolonged period of cold before flowering occurs.

History of Vernalization Research

In the history of agriculture, farmers observed a traditional distinction between "winter cereals," whose seeds require chilling, and "spring cereals," whose seeds can be sown in spring and flower soon thereafter. The word "vernalization" is a translation of "ÿðîâèçàöèÿ" ("jarovization") a word coined by Trofim Lysenko to describe a chilling process he used to make the seeds of winter cereals behave like spring cereals ("Jarovoe" in Russian). Scientists had also discussed how some plants needed cold temperatures to flower, as early as the 18th century, with the German plant physiologist Gustav Gassner often mentioned for his 1918 paper.

Lysenko's 1928 paper on vernalization and plant physiology drew wide attention due to its practical consequences for Russian agriculture. Severe cold and lack of winter snow had destroyed many early winter wheat seedlings. By treating wheat seeds with moisture as well as cold, Lysenko induced them to bear a crop when planted in spring. Later however, Lysenko inaccurately asserted that the vernalized state could be inherited - i.e., that the offspring of a vernalized plant would behave as if they themselves had also been vernalized and would not require vernalization in order to flower quickly.

Early research on vernalization focused on plant physiology; the increasing availability of molecular biology has made it possible to unravel its underlying mechanisms. For example, longer days as well as cold temperatures are required for winter wheat plants to go from the vegetative to the reproductive state; the three interacting genes are called *VRN1*, *VRN2*, and *FT* (*VRN3*).

Due to plant flowering requiring the successful co-operation of several metabolic pathways, computer models that incorporate vernalization have also been made.

Vernalization in Arabidopsis Thaliana

Arabidopsis thaliana, also known as "thale cress," is a much-studied model species. In 2000, the entire genome of its five chromosomes was completely sequenced. Some variants, called "winter annuals", require vernalization to flower; others ("summer annuals") do not. The genes that underlie this difference in plant physiology have been intensively studied.

The reproductive phase change of *A. thalliana* occurs by a sequence of two related events: first, the bolting transition (flower stalk elongates), then the floral transition (first flower appears). Bolting is a robust predictor of flower formation, and hence a good indicator for vernalization research. In arabidopsis winter annuals, the apical meristem is the part of the plant that needs to be chilled. Vernalization of the meristem appears to confer competence to respond to floral inductive signals on the meristem. A vernalized meristem retains competence for as long as 300 days in the absence of an inductive signal. Before vernalization, flowering is repressed by the action of a gene called *Flowering Locus C* (*FLC*). Prolonged exposure to cold induces expression of VERNALIZATION INSENSTIVE3, which interacts with the VERNALIZATION2 polycomb-like complex to reduce FLC expression through chromatin remodelling. Since vernalization also occurs in *flc* mutants (lacking *FLC*), vernalization must also activate a non-*FLC* pathway. A day-length mechanism is also important.

Devernalization

It is possible to devernalize a plant by exposure to high temperatures subsequent to vernalization. For example, commercial onion growers store seeds at low temperatures, but devernalize them before planting, because they want the plant's energy to go into enlarging its bulb (underground stem), not making flowers.

Antifreeze Protein

Antifreeze proteins (AFPs) or ice structuring proteins (ISPs) refer to a class of polypeptides produced by certain vertebrates, plants, fungi and bacteria that permit their survival in subzero environments.

AFPs bind to small ice crystals to inhibit growth and recrystallization of ice that would otherwise be fatal. There is also increasing evidence that AFPs interact with mammalian cell membranes to protect them from cold damage. This work suggests the involvement of AFPs in cold acclimatization.

Non-Colligative Properties

Unlike the widely used automotive antifreeze, ethylene glycol, AFPs do not lower freezing point in proportion to concentration. Rather, they work in a noncolligative manner. This phenomenon allows them to act as an antifreeze at concentrations 1/300th to 1/500th of those of other dissolved solutes. Their low concentration minimizes their effect on osmotic pressure. The unusual properties of AFPs are attributed to their affinity for specific ice crystal surfaces.

Thermal Hysteresis

AFPs create a difference between the melting point and freezing point known as thermal hysteresis. The addition of AFPs at the interface between solid ice and liquid water inhibits the thermodynamically favoured growth of the ice crystal. Ice growth is kinetically inhibited by the AFPs covering the water-accessible surfaces of ice.

Thermal hysteresis is easily measured in the lab with a nanolitre osmometre. Organisms differ in their values of thermal hysteresis. The maximum level of thermal hysteresis shown by fish AFP is approximately -1.5°C (29.3°F). However, insect antifreeze proteins are 10–30 times more active than fish proteins. This difference probably reflects the lower temperatures encountered by insects on land. In contrast, aquatic organisms are exposed only to –1 to –2°C below freezing. During the extreme winter months, the spruce budworm resist freezing at temperatures approaching –30 °C. The Alaskan beetle *Upis ceramboides* can survive in a temperature of –60 °C by using antifreeze agents that are not proteins.

The rate of cooling can influence the thermal hysteresis value of AFPs. Rapid cooling can substantially decrease the nonequilibrium freezing point, and hence the thermal hysteresis value. Consequently, organisms cannot necessarily adapt to their subzero environment if the temperature drops abruptly.

Freeze Tolerance versus Freeze Avoidance

Species containing AFPs may be classified as

Freeze avoidant: These species are able to prevent their body fluids from freezing altogether. Generally, the AFP function may be overcome at extremely cold temperatures, leading to rapid ice growth and death.

Freeze tolerant: These species are able to survive body fluid freezing. Some freeze tolerant species are thought to use AFPs as cryoprotectants to prevent the damage of freezing, but not freezing altogether. The exact mechanism is still unknown. However, it is thought AFPs may inhibit recrystallization and stabilize cell membranes to prevent damage by ice. They may work in conjunction with protein ice nucleators (PINs) to control the rate of ice propagation following freezing.

Diversity

There are many known nonhomologous types of AFPs.

Fish AFPs

Antifreeze glycoproteins or AFGPs are found in Antarctic notothenioids and northern cod. They are 2.6-3.3 kD. AFGPs evolved separately in notothenioids and northern cod. In notothenioids, the AFGP gene arose from an ancestral trypsinogen-like serine protease gene.

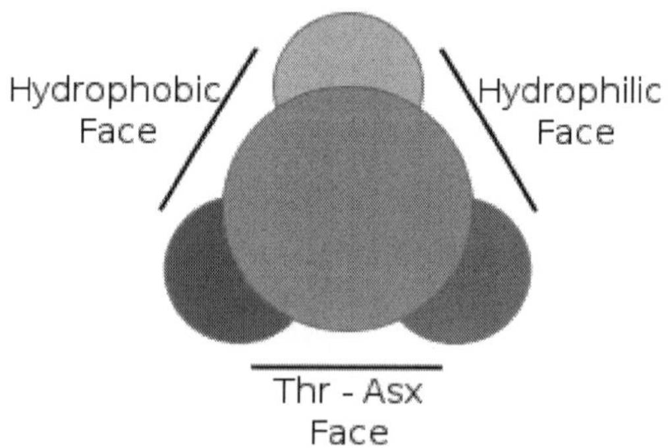

Figure: *The three faces of Type I AFP*

Type I AFP is found in winter flounder, longhorn sculpin and shorthorn sculpin. It is the best documented AFP because it was the first to have its three-dimensional structure determined. Type I AFP consists of a single, long, amphipathic alpha helix, about 3.3-4.5 kD in size. There are three faces to the 3D structure: the hydrophobic, hydrophilic, and Thr-Asx face.

Type I-hyp AFP (where hyp stands for hyperactive) are found in several righteye flounders. It is approximately 32 kD (two 17 kD dimeric molecules). The protein was isolated from the blood plasma of winter flounder. It is considerably better at depressing freezing temperature than most fish AFPs.

Type II AFPs are found in sea raven, smelt and herring. They are cysteine-rich globular proteins containing five disulfide bonds. Type II AFPs likely evolved from calcium dependent (c-type) lectins. Sea ravens, smelt, and herring are quite divergent lineages of teleost. If the AFP gene was present in the most recent common ancestor of these lineages, it's peculiar that the gene is scattered throughout those lineages, present in some orders and absent in others. It has been suggested that lateral gene transfer could be attributed to this discrepancy, such that the smelt acquired the type II AFP gene from the herring.

Type III AFPs are found in Antarctic eelpout. They exhibit similar overall hydrophobicity at ice binding surfaces to type I AFPs. They are approximately 6kD in size. Type III AFPs likely evolved from a sialic acid synthase gene present in Antarctic zoarcid fish. Through a gene duplication event, this gene—which has been shown to exhibit some ice-binding activity of its own—evolved into an effective AFP gene.

Type IV AFPs are found in longhorn sculpins. They are alpha helical proteins rich in glutamate and glutamine. This protein is approximately 12KDa in size and consists of a 4-helix bundle. Its only posttranslational modification is a pyroglutamate residue, a cyclized glutamine residue at its N-terminus. Scientists at the University of Guelph in Canada are currently examining the role of this pyroglutame residue in the antifreeze activity of type IV AFP from the longhorn sculpin.

Plant AFPs

The classification of AFPs became more complicated when antifreeze proteins from plants were discovered. Plant AFPs are rather different from the other AFPs in the following aspects:

1. They have much weaker thermal hysteresis activity when compared to other AFPs.
2. Their physiological function is likely in inhibiting the recrystallization of ice rather than in the preventing ice formation.
3. Most of them are evolved pathogenesis-related proteins, sometimes retaining antifungal properties.

Insect AFPs

There are two types of insect antifreeze proteins, *Tenebrio* and *Dendroides* AFPs which are both in different insect families. They are similar to one another, both being hyperactive (i.e. greater thermal hysteresis value) and consist of varying numbers of 12- or 13-mer repeats of approximately 8.3 to 12.5 kD. Throughout the length of the protein, at least every sixth residue is a cysteine.

Tenebrio or Type V AFPs are found in beetles, whereas *Dendroides* or *Choristoneura fumiferana* AFPs are found in some Lepidoptera.

Sea Ice Organisms AFPs

AFPs were also found in microorganisms living in sea ice. The diatoms *Fragilariopsis cylindrus* and *F. curta* play a key role in polar sea ice communities, dominating the assemblages of both platelet layer and within pack ice. AFPs are widespread in these species, and the presence of AFP genes as a multigene family indicates the importance of this group for the genus *Fragilariopsis*. AFPs identified in *F. cylindrus* belong to an AFP family which is represented in different taxa and can be found in other organisms related to sea ice (*Colwellia* spp., *Navicula glaciei, Chaetoceros neogracile* and *Stephos longipes and Leucosporidium antarcticum*) and Antarctic inland ice bacteria (Flavobacteriaceae), as well as in cold-tolerant fungi (*Typhula ishikariensis, Lentinula edodes* and *Flammulina populicola*.)

Evolution

The remarkable diversity and distribution of AFPs suggest the different types evolved recently in response to sea level glaciation occurring 1-2 million years ago in the Northern hemisphere and 10-30 million years ago in Antarctica. This independent development of similar adaptations is referred to as convergent evolution. There are two reasons why many types of AFPs are able to carry out the same function despite their diversity:

1. Although ice is uniformly composed of oxygen and hydrogen, it has many different surfaces exposed for binding. Different types of AFPs may interact with different surfaces.
2. Although the five types of AFPs differ in their primary sequence of amino acids, when each folds into a functioning protein, they may share similarities in their three-dimensional or tertiary structure that facilitates the same interactions with ice.

Mechanisms of Action

AFPs are thought to inhibit growth by an adsorption–inhibition mechanism. They adsorb to nonbasal planes of ice, inhibiting thermodynamically favoured ice growth. The presence of a flat, rigid surface in some AFPs seems to facilitate its interaction with ice via Van der Waals force surface complementarity.

Binding to Ice

Normally, ice crystals grown in solution only exhibit the basal (0001) and prism faces (1010), and appear as round and flat discs. However, it appears the presence of AFPs exposes other faces. It now appears the ice surface 2021 is the preferred binding surface, at least for AFP type I. Through studies on type I AFP, ice and AFP were initially thought to interact through hydrogen bonding (Raymond and DeVries, 1977). However, when parts of the protein thought to facilitate this hydrogen bonding were mutated, the hypothesized decrease in antifreeze activity was not observed. Recent data suggest hydrophobic interactions could be the main contributor. It is difficult to discern the exact mechanism of binding because of the complex water-ice interface. Currently, attempts to uncover the precise mechanism are being made through use of molecular modelling programs (molecular dynamics or the Monte Carlo method).

Binding Mechanism and Antifreeze Function

According to the structure and function study on the antifreeze protein from the fish winter flounder, the antifreeze mechanism of the type-I AFP molecule was shown to be due to the binding to an ice nucleation structure in a zipper-like fashion through hydrogen bonding of the hydroxyl groups of its four Thr residues to the oxygens along the $[01\overline{1}2]$ direction in ice lattice, subsequently stopping or retarding the growth of ice pyramidal planes so as to depress the freeze point. The above mechanism can be used to elucidate the structure-function relationship of other antifreeze proteins with the following two common features:

1. recurrence of a Thr residue (or any other polar amino acid residue whose side-chain can form a hydrogen bond with water) in an 11-amino-acid period along the sequence concerned, and
2. a high percentage of an Ala residue component therein.

History

In the 1950s, Canadian scientist Scholander set out to explain how Arctic fish can survive in water colder than the freezing point of their blood. His experiments led him to believe there was "antifreeze" in the blood of Arctic fish. Then in the late 1960s, animal biologist Arthur DeVries was able to isolate the antifreeze protein through his investigation of Antarctic fish. These proteins were later called antifreeze glycoproteins (AFGPs) or antifreeze glycopeptides to distinguish them from newly discovered nonglycoprotein biological antifreeze agents (AFPs). DeVries worked with Robert Feeney (1970) to characterize the chemical and physical properties of antifreeze proteins. In 1992, Griffith *et al.* documented their discovery of AFP in winter rye leaves. Around the same time, Urrutia, Duman and Knight (1992) documented thermal hysteresis protein in angiosperms. The next year, Duman and Olsen noted AFPs had also been discovered in over 23 species of angiosperms, including ones eaten by humans. As well, they reported their presence in fungi and bacteria.

Name Change

Recent attempts have been made to relabel antifreeze proteins as ice structuring proteins to more accurately represent their function and to dispose of any assumed negative relation between AFPs and automotive antifreeze, ethylene glycol. These two things are completely separate entities, and show loose similarity only in their function.

Commercial Applications

Commercially, there appear to be countless applications for antifreeze proteins. Numerous fields would be able to benefit from the protection of tissue damage by freezing. Businesses are currently investigating the use of these proteins in:

- increasing freeze tolerance of crop plants and extending the harvest season in cooler climates
- improving farm fish production in cooler climates
- lengthening shelf life of frozen foods
- improving cryosurgery
- enhancing preservation of tissues for transplant or transfusion in medicine
- therapy for hypothermia

Dehydrin

Dehydrin (DHN) is a multi-family of proteins present in plants that is produced in response to cold and drought stress. DHNs are hydrophilic and reliably thermostable. They are stress proteins with a high number of charged amino acids that belong to the Group II Late Embryogenesis Abundant (LEA) family". DHNs are primarily found in the cytoplasm and nucleus but more recently, they have been found in other organelles, like mitochondria and chloroplasts. DHNs are characterized by the presence of Glycine and other polar amino acids. All DHNs contain at least one copy of a consensus 15-amino acid sequence. The "K segment, The K segment is a Lysine-rich 15-amino acid consensus sequence (EKKGIMDKIKEKPLG) that is highly conserved in all plants".

Dehydration-induced proteins in plants were first observed in 1989, in a comparison of barley and corn cDNA from plants under drought conditions. The protein has since been referred to as dehydrin and has been the identified as the genetic basis of drought tolerance in plants. However, the first direct genetic evidence of dehydrin playing a role in cellular protection during osmotic shock was not observed until 2005, in the moss, Physcomitrella patens. In order to show a direct correlation between DHN and stress recovery, a knockout gene was created, which interfered with DHNA's functionality. After being placed in an environment with salt and osmotic stress and then later being returned to a standard growth medium, the P. patens wildtype was able to recover to 94% of its fresh weight while the P. patens mutant only reached 39% of its fresh weight. This study also concludes that DHN production allows plants to function in high salt concentrations. Another study found evidence of DHN's impact in drought-stress recovery by showing that transcription levels of a DHN increased in a drought-tolerant pine, Pinus pisaster, when placed in a drought treatment. However, transcription levels of a DHN decreased in the same drought treatment in a drought-sensitive P. pisaster. Drought-tolerance is a complex trait, thus that it cannot be genetically analyzed as a single gene trait. The exact mechanism of drought tolerance is yet to be determined and is still being researched. One chemical mechanism related to DHN production is the presence of the phytohormone ABA. One common response to environmental stresses is process known as cellular dehydration. Cellular dehydration induces biosynthesis of abscisic acid (ABA), which is known to react as a stress hormone because of its accumulation in the plant under water stress conditions. ABA also participates in stress signal transduction pathways ABA has been shown to increase the production of DHN, which provides more evidence of a link between DHN and drought tolerance.

There are other proteins in the cell that play a similar role in the recovery of drought treated plants. These proteins are considered dehydrin-like or dehydrin-related. They are poorly defined, in that these dehydrin-like proteins are similar to DHNs, but are unfit to be classified as DHNs for varying reasons. They are found to be similar in that they respond to some or all of the same environmental stresses that induce DHN production. In a particular study dehydrin-like proteins found in the mitochondria were upregulated in drought and cold treatments of cereals.

10

Staying Stiff and High

As can be seen from Figure earlier, the tissues comprise an outer epidermis (made up of a single layer of closely packed cells), vascular bundles and parenchyma. Parenchyma cells, which make up the bulk of the stem, are thin walled with large vacuoles. In leaves and stems of seedlings and small plants it is the water content of these cells that holds the plant erect.

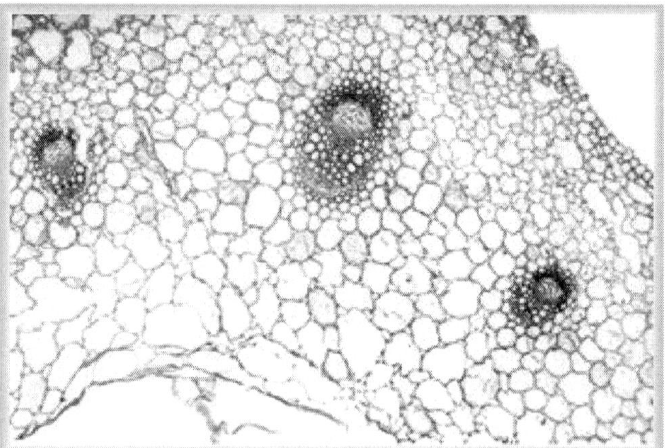

Figure: *TS* Ranuculus *stem*

A plant cell is different from an animal cell in that it has a cellulose cell wall and also in that it contains a large sap filled vacuole. If plant cells are placed in a solution which has a weaker solute concentration than the cell sap, then water enters the cell vacuole by osmosis. This causes the protoplast to be pushed against the cell wall exerting pressure against it and preventing the cell from bursting. Cells in this state are said to be turgid. It is this cell turgor that gives the plant support. If however, plant cells are surrounded by a solution with a stronger solute concentration than the cell sap, then the vacuole loses water and as a result there is no hydrostatic pressure acting against the cell wall, such cells are said to be flaccid or plasmolysed. Even those of us who lack green fingers are aware that a drooping, wilting plant has not been watered; once the necessary watering has been carried out the stems and leaves very quickly revert to their erect state as parenchyma cells become turgid.

Turgid and plasmolysed cells can be seen when pieces of the thin membrane from a red onion are placed in different solutions of different concentrations.

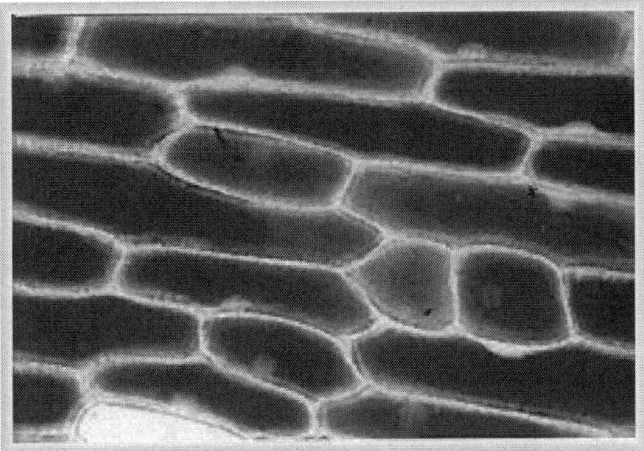

Figure: *Turgid cells*

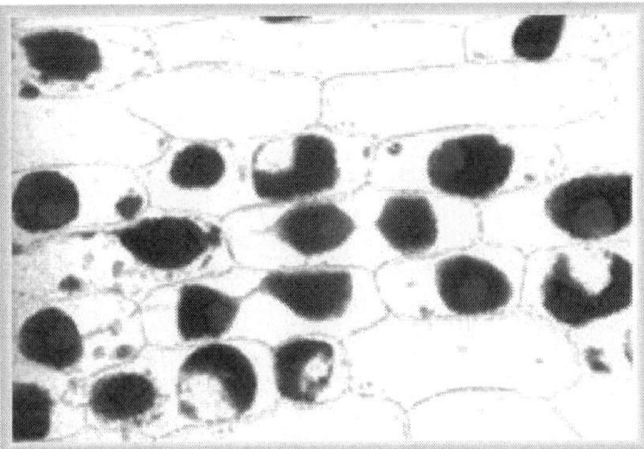

Figure: *Plasmolysed cells*

While turgor and osmosis play an important role, other factors are also involved. When transverse sections of dicotyledons stems are observed, vascular tissue arranged in vascular bundles can be seen situated near the periphery of the stem. Vascular tissue is involved with the transport of material. The vascular bundles are made up of phloem (closest to the outside), xylem, with cambium between. Phloem is a specialised form of parenchyma, made up of sieve elements and companion cells, whose function is the transport of materials produced during photosynthesis. Fibres within the phloem tissue help in supporting the plant.

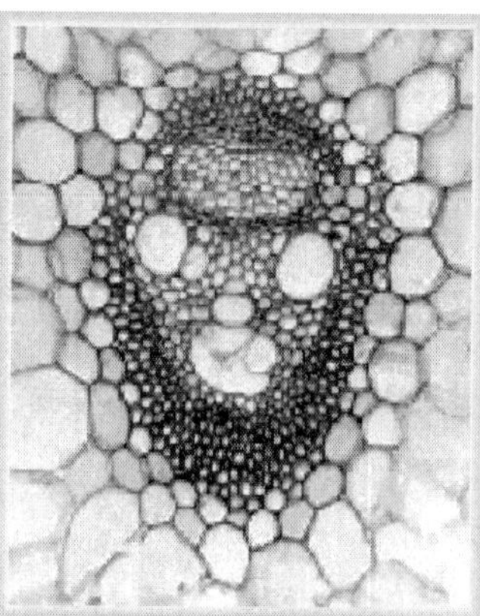

Figure: *Vascular bundle*

Xylem, which does not contain living material, transports water and dissolved mineral salts from the roots and also plays an important role in support. Xylem tissue is made up of tracheary elements, which conduct water and dissolved minerals, together with thick non conducting fibres. The hollow tubes of xylem tissue are usually strengthened with lignin, the woody material we are familiar with in trees and shrubs. The strengthening is in the form of hoops or spirals of lignin as can be seen in the following figure showing lignified xylem vessels in a transverse section of rhubarb (*Rheum*)

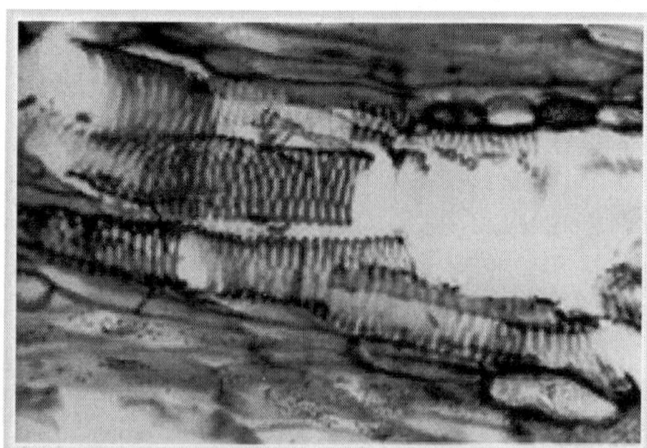

Figure: Rheum *Xylem vessels showing lignification*

Vascular bundles in dicotyledons can be seen to be arranged near to the outside of the stem; often too, the stem may be hollow. This cylindrical formation provides greater strength than a solid structure of the same weight. Vascular arrangement in the root is different, phloem and xylem are found in a central stele.

This can be seen in Figure given, showing a transverse section through a *Ranunculus* root. The xylem vessels forming the X shaped structure, with phloem inbetween the arms of the X. The vascular tissue is enclosed by a layer of cells called the endodermis. This central arrangement of vascular tissue provides strength against the pulling forces that the root may be subjected to. This same arrangement can also sometimes be seen in the stems of aquatic plants which are subjected to different forces than the stems of their terrestrial relations.

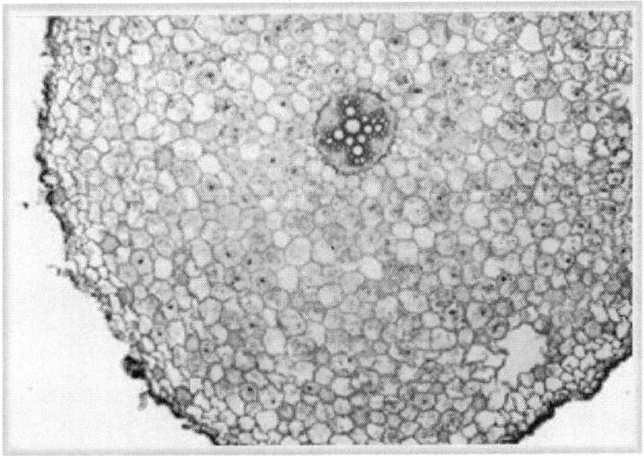

Figure: *TS* Ranuculus *root*

Parenchyma, phloem and xylem are not the only tissues visible in a stem cross section; collenchyma often located in the outer tissue has thick cell walls and is relatively pliable, it can play a significant part in the support of shoots but is seldom found in roots. It is often found at the angles of the stem as can be seen in the following picture of a transverse section of a *Lamium* stem.

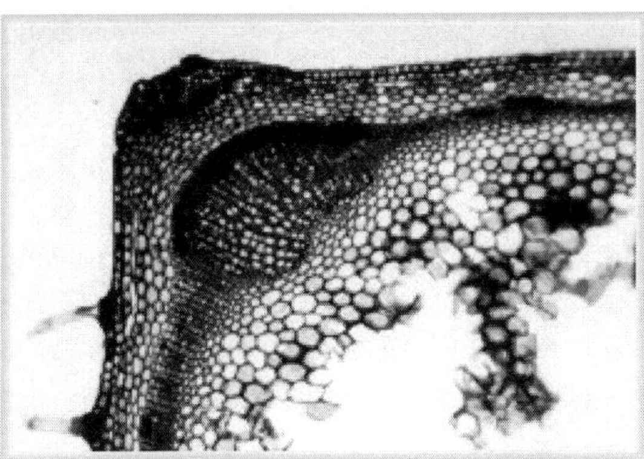

Figure: *TS* Lamium *stem*

The cell walls of collenchyma cells are rich in pectic substances; fibrils within the cell wall are arranged such that the cell wall has great plasticity, letting the cells to deform without damage and so allowing the stem to bend without breaking. Often scattered among the cells of the tissues mentioned is a further type of cell, sclerenchyma. This tissue is characterised by having a secondary cell wall (inside the primary cell wall) which gives the cell elasticity, allowing it to return to its original shape after being bent. The angle of the fibrils within the secondary cell wall determines the stiffness of the cell. The arrangement of the fibrils is often varied to confer strength and stability against a range of forces.

Plants Exhibit a Wide Range of Mechanical Properties, Engineers Find

From an engineer's perspective, plants such as palm trees, bamboo, maples and even potatoes are examples of precise engineering on a microscopic scale. Like wooden beams reinforcing a house, cell walls make up the structural supports of all plants. Depending on how the cell walls are arranged, and what they are made of, a plant can be as flimsy as a reed, or as sturdy as an oak.

An MIT researcher has compiled data on the microstructures of a number of different plants, from apples and potatoes to willow and spruce trees, and has found that plants exhibit an enormous range of mechanical properties, depending on the arrangement of a cell wall's four main building blocks: cellulose, hemicellulose, lignin and pectin.

Lorna Gibson, the Matoula S. Salapatas Professor of Materials Science and Engineering at MIT, says understanding plants' microscopic organisation may help engineers design new, bio-inspired materials.

"If you look at engineering materials, we have lots of different types, thousands of materials that have more or less the same range of properties as plants," Gibson says. "But here the plants are, doing it arranging just four basic constituents. So maybe there's something you can learn about the design of engineered materials."

A paper detailing Gibson's findings has been published this month in the *Journal of the Royal Society Interface.*

To Gibson, a cell wall's components bear a close resemblance to certain manmade materials. For example, cellulose, hemicellulose and lignin can be as stiff and strong as manufactured polymers. A plant's cellular arrangement can also have engineering parallels: cells in woods, for instance, are aligned, similar to engineering honeycombs, while polyhedral cell configurations, such as those found in apples, resemble some industrial foams.

To explore plants' natural mechanics, Gibson focused on three main plant materials: woods, such as cedar and oak; parenchyma cells, which are found in fruits and root vegetables; and arborescent palm stems, such as coconut trees. She compiled data from her own and other groups' experiments and analyzed two main mechanical properties in each plant: stiffness and strength.

Among all plants, Gibson observed wide variety in both properties. Fruits and vegetables such as apples and potatoes were the least stiff, while the densest palms were 100,000 times stiffer. Likewise, apples and potatoes fell on the lower end of the strength scale, while palms were 1,000 times stronger.

"There are plants with properties over that whole range," Gibson says. "So it's not like potatoes are down here, and wood is over there, and there's nothing in between. There are plants with properties spanning that whole huge range. And it's interesting how the plants do that."

It turns out the large range in stiffness and strength stems from an intricate combination of plant microstructures: the composition of the cell wall, the number of layers in the cell wall, the arrangement of cellulose fibers in those layers, and how much space the cell wall takes up.

In trees such as maples and oaks, cells grow and multiply in the cambium layer, just below the bark, increasing the diameter of the trees. The cell walls in wood are composed of a primary layer with cellulose fibers randomly spread throughout it. Three secondary layers lie underneath, each with varying compositions of lignin and cellulose that wind helically through each layer.

Taken together, the cell walls occupy a large portion of a cell, providing structural support. The cells in woods are organised in a honeycomb pattern — a geometric arrangement that gives wood its stiffness and strength.

Parenchyma cells, found in fruits and root vegetables, are much less stiff and strong than wood. The cell walls of apples, potatoes and carrots are much thinner than in wood cells, and made up of only one layer. Cellulose fibers run randomly throughout this layer, reinforcing a matrix of hemicellulose and pectin. Parenchyma cells have no lignin;

combined with their thin walls and the random arrangement of their cellulose fibers, Gibson says, this may explain their cell walls' low stiffness. The cells in each plant are densely packed together, similar to industrial foams used in mattresses and packaging.

Unlike woody trees that grow in diameter over time, the stems of arborescent palms such as coconut trees maintain similar diameters throughout their lifetimes. Instead, as the stem grows taller, palms support this extra weight by increasing the thickness of their cell walls. A cell wall's thickness depends on where it is along a given palm stem: Cell walls are thicker at the base and periphery of stems, where bending stresses are greatest.

Gibson sees plant mechanics as a valuable resource for engineers designing new materials. For instance, she says, researchers have developed a wide array of materials, from soft elastomers to stiff, strong alloys. Carbon nanotubes have been used to reinforce composite materials, and engineers have made honeycomb-patterned materials with cells as small as a few millimetres wide. But researchers have been unable to fabricate cellular composite materials with the level of control that plants have perfected.

"Plants are multifunctional," Gibson says. "They have to satisfy a number of requirements: mechanical ones, but also growth, surface area for sunlight and transport of fluids.

The microstructures plants have developed satisfy all these requirements. With the development of nanotechnology, I think there is potential to develop multifunctional engineering materials inspired by plant microstructures."

Karl Niklas, a professor of plant biology at Cornell University, says Gibson's engineering parallels are fitting. Plants, in a way, he says, are "largely structural things ... chemical factories that are architecturally arranged."

"Plants on Earth have evolved over three-and-a-half billion years, and that is a giant evolutionary experiment of trial and error, because the things that don't work are extinct, and the things that do work are more abundant," Niklas says. "We can learn things from nature and apply it to construct better panel boards, styrofoams and photovoltaics that will help society."

Which Plants will Survive Droughts, Climate Change?

New research by UCLA life scientists could lead to predictions of which plant species will escape extinction from climate change.

Droughts are worsening around the world, posing a great challenge to plants in all ecosystems, said Lawren Sack, a UCLA professor of ecology and evolutionary biology and senior author of the research. Scientists have debated for more than a century how to predict which species are most vulnerable.

Sack and two members of his laboratory have made a fundamental discovery that resolves this debate and allows for the prediction of how diverse plant species and vegetation types worldwide will tolerate drought, which is critical given the threats posed by climate change, he said. The research is currently available in the online edition of Ecology Letters, a prestigious ecology journal, and will be published in an upcoming print edition.

Why does a sunflower wilt and dessicate quickly when the soil dries, while the native chaparral shrubs of California survive long dry seasons with their evergreen leaves? Since there are many mechanisms involved in determining the drought tolerance of plants, there has been vigorous debate among plant scientists over which trait is most important. The UCLA team, funded by the National Science Foundation, focused on a trait called "turgor loss point, which had never before been proven to predict drought tolerance across plant species and ecosystems.

A fundamental difference between plants and animals is that plant cells are enclosed by cell walls while animal cells are not. To keep their cells functional, plants depend on "turgor pressure" — pressure produced in cells by internal salty water pushing against and holding up the cell walls. When leaves open their pores, or stomata, to capture carbon dioxide for photosynthesis, they lose a considerable amount of this water to evaporation. This dehydrates the cells, inducing a loss of pressure.

During drought, the cell's water becomes harder to replace. The turgor loss point is reached when leaf cells get to a point at which their walls become flaccid; this cell-level loss of turgor causes the leaf to become limp and wilted, and the plant cannot grow, Sack said.

"Drying soil may cause a plant's cells to reach turgor loss point, and the plant will be faced with the choice of either closing its stomata and risking starvation or photosynthesizing with wilted leaves and risking damaging its cell walls and metabolic proteins," Sack said. "To be more drought-tolerant, the plant needs to change its turgor loss point so that its cells will be able to keep their turgor even when soil is dry."

The biologists showed that within ecosystems and around the world, plants that are more drought-tolerant had lower turgor loss points; they could maintain their turgor despite drier soil.

The team also resolved additional decades-old controversies, overturning the long-held assumptions of many scientists about the traits that determine turgor loss point and drought tolerance. Two traits related to plant cells have been thought to affect plants' turgor loss point and improve drought tolerance: Plants can make their cell walls stiffer or they can make their cells saltier by loading them with dissolved solutes.

Many prominent scientists have leaned toward the "stiff cell wall" explanation because plants in dry zones around the globe tend to have small, tough leaves. Stiff cell walls might allow the leaf to avoid wilting and to hold onto its water during dry times, scientists reasoned. Little had been known about the saltiness of cells for plants around the world.

The UCLA team has now demonstrated conclusively that it is the saltiness of the cell sap that explains drought tolerance across species. Their first approach was mathematical; the team revisited the fundamental equations that govern wilting behaviour and solved them for the first time.

Their mathematical solution pointed to the importance of saltier cell sap. Saltier cell sap in each plant cell allows the plant to maintain turgor pressure during dry times and to continue photosynthesizing and growing as drought ensues. The equation showed that thick cell walls do not contribute directly to preventing wilting, although they provide indirect benefits that can be important in some cases — protection from excessive cell shrinking and from damage due to the elements or insects and mammals.

The team also collected for the first time drought-tolerance trait data for species worldwide, which confirmed their result. Across species within geographic areas and across the globe, drought tolerance was correlated with the saltiness of the cell sap and not with the stiffness of cell walls. In fact, species with stiff cell walls were found not only in arid zones but also in wet systems like rainforests, because here too, evolution favours long-lived leaves protected from damage.

The pinpointing of cell saltiness as the main driver of drought tolerance cleared away major controversies, and it opens the way to predictions of which species could escape extinction from climate change, Sack said.

"The salt concentrated in cells holds on to water more tightly and directly allows plants to maintain turgor during drought," said research co-author Christine Scoffoni, a UCLA doctoral student in the department of ecology and evolutionary biology.

The role of the stiff cell wall was more elusive.

"We were surprised to see that having a stiffer cell wall actually *reduced* drought tolerance slightly — contrary to received wisdom — but that many drought-tolerant plants with lots of salt also had stiff cell walls," said lead author Megan Bartlett, a UCLA graduate student in the department of ecology and evolutionary biology.

This seeming contradiction is explained by the secondary need of drought-tolerant plants to protect their dehydrating cells from shrinking as they lose turgor pressure, the researchers said.

"While a stiff wall doesn't maintain the cell turgor, it prevents the cells from shrinking as the turgor decreases and holds in water so that cells are still large and hydrated, even at turgor loss point," Bartlett explained.

"So the ideal combination for a plant is to have a high solute concentration to keep turgor pressure and a stiff cell wall to prevent it from losing too much water and shrinking as the leaf water pressure drops. But even drought-sensitive plants often have thick cell walls because the tough leaves are also good protection against herbivores and everyday wear and tear."

Even though the team showed that turgor loss point and salty cell sap have exceptional power to predict a plant's drought tolerance, some of the most famous and diverse desert plants — including cacti, yuccas and agaves — exhibit the opposite design, with many flexible-walled cells that hold dilute sap and would lose turgor rapidly, Sack said.

"These succulents are actually terrible at tolerating drought, and instead they avoid it," he said. "Because much of their tissue is water storage cells, they can open their stomata minimally during the day or at night and survive with their stored water until it rains. Flexible cell walls help them release water to the rest of the plant."

This new study showed that the saltiness of cells in plant leaves can explain where plants live and the kinds of plants that dominate ecosystems around the world.

The team is working with collaborators at the Xishuangbanna Tropical Botanical Gardens in Yunnan, China, to develop a new method for rapidly measuring turgor loss point across a large number of species and make possible the critical assessment of drought tolerance for thousands of species for the first time.

"We're excited to have such a powerful drought indicator that we can measure easily," Bartlett said. "We can apply this across entire ecosystems or plant families to see how plants have adapted to their environment and to develop better strategies for their conservation in the face of climate change."

11

Surviving a Storm

Understanding Lightning and Associated Tree Damage

Lightning is one of nature's most powerful forces. Lightning can have devastating effects on people, property and trees. Each strike of lightning can reach more than five miles in length, and produce temperatures greater than 50,000 degrees Fahrenheit and an electrical charge of 100 million volts. At any given moment, there are 1,800 thunderstorms in progress somewhere across the earth. Lightning detection systems in the United States sense an average of 25 million lightning strikes per year.

Trees occupy a particularly susceptible position in the landscape, since they are often the tallest objects. Tall trees are the most vulnerable, especially those growing alone in open areas such as on hills, in pastures or near water. Many of these trees line our community streets and surround our homes, schools and businesses.

Response of Trees to Lightening

A tree's biological functions and/or structural integrity are affected by lightning strikes. Along the path of the strike, sap boils, steam is generated and cells explode in the wood, leading to strips of wood and bark peeling or being blown off the tree. If only one side of the tree shows evidence of a lightning strike, the chances of the tree surviving and eventually closing the wound are good. However, when the strike completely passes through the tree trunk, with splintered bark and exploded wood on each side, trees are usually killed.

Many trees are severely injured internally or below- ground by lightning despite the absence of visible, external symptoms. Lightning or electrical current passes from the trunk of the tree through the roots and dissipates in the ground. Major root damage from electricity may cause the tree to decline and die without significant aboveground damage. If the tree is in leaf, the leaves wilt and the tree will probably die within a few days. If the tree survives long enough to leaf out the following spring, then the chances of recovery are much greater. Watering and fertilization are suggested to reduce tree stress.

Generally, when lightning damage has created hazardous broken branches, corrective pruning should be done. However, waiting two to six months is recommended before doing major and expensive corrective pruning to assess whether the tree will recover. If during this waiting period, the tree shows no obvious signs of decline, then the pruning is probably worth the expense. Consult with a certified arborist for recommendations concerning the health of your damaged tree. Commonly prescribed practices are water management, bark repair, pruning, fertilization, pest management and tree monitoring. Expensive treatments should not be taken until the tree appears to be making a recovery. Otherwise, when it becomes obvious that the tree will not recover from the lightning strike, the tree should be removed.

Lightning Protection Systems for Trees

Wayne Clatterbuck

Lightning scar on an eastern white pine tree.

Historic, rare, and specimen trees, especially when they are the centre of landscapes or they shade or frame recreational areas, are valuable and can be protected by a properly installed lightning protection system. Trees with special significance, or that people or animals might move under in a storm, should be protected. Trees closer than 25 feet from a building or structure should also be protected to minimize "side-flash." Parks, golf courses, and public buildings should have large or important trees protected to minimize liability risks.

Tree lightning protection is expensive in labour and materials. Lightning protection systems must be installed properly with correct materials to insure long-term protection. For example, aluminum should not be used for any link in a system, nor should solid wire of any type be used. It is essential to consult with a trained arborist or urban forester, and a lightning protection system installer before designing a protection system for a tree.

Lightning protection systems in trees do not attract lightning. The purpose of a protection system is to dilute and slowly release electrical charge potential between the ground and cloud. Trees are not good conductors of electricity but can act as a better conduit than air. Protection systems dissipate the electrical charge before it can build to high levels.

Summary

Lightning damage in trees is much more effectively prevented than repaired and is often less costly. Qualified arborists can recommend the installation of lightning protection systems where appropriate or the proper course of action if a tree has been struck by lightning.

Can Leaves Predict a Storm?

When leaves show their undersides, be very sure rain betides."

Weather folklore has always been a big part of the *Farmers' Almanac*. Long before there were scientific instruments to measure and predict the weather, people used the only instrument they had: the power of observation.

"Old wives' tales," like the one recorded above came about because people noticed patterns in nature, and passed those observations down from generation to generation. Whereas today, we can just switch on the local news (or crack open a *Farmers' Almanac*) to get a weather report, our ancestors had to rely on their senses.

But does seeing the undersides of leaves really mean rain is on the way? In this case, our forebears were definitely onto something. The leaves of deciduous trees, like maples and poplars, do often to turn upward before heavy rain. The leaves are actually reacting to the sudden increase in humidity that usually precedes a storm. Leaves with soft stems can become limp in response to abrupt changes in humidity, allowing the wind to flip them over.

Why do Leaves Turn Upside Down before a Storm?

Leaves turn upside down before a storm because of the change in conditions in the atmosphere as warm or cold fronts come through. The change in temperature causes cooler air to drop and warm air to rise, creating a breeze. Humidity makes the leaves more pliant and they are more easily flipped from the updraft of the breezes leading up to the storm.

Trees and Our Climate

Trees affect our climate, and therefore our weather, in three primary ways: they lower temperatures, reduce energy usage and reduce or remove air pollutants. Each part of the tree contributes to climate control, from leaves to roots.

Leaves help turn down the thermostat. They cool the air through a process called evapotranspiration. Evapotranspiration is the combination of two simultaneous processes: evaporation and transpiration, both of which release moisture into the air. During evaporation, water is converted from liquid to vapor and evaporates from soil, lakes, rivers and even pavement. During transpiration, water that was drawn up through the soil by the roots evaporates from the leaves. It may seem like an invisible process to our eyes, but a large oak tree is capable of transpiring 40,000 gallons of water into the atmosphere during one year.

The outdoor air conditioning provided by trees reduces the energy used inside your home or office. Shade provided by strategically planted deciduous trees cools buildings during the warm months, allows the sun's warming rays to shine through its branches in the winter and also protects buildings from cold winds. With some planning, urban trees can help minimize the heat island effect that saddles many cities.

Heat islands are cities that are often several degrees warmer than the suburbs because the urban areas generate and trap heat. Studies of Atlanta found that temperatures downtown were 5 to 8 degrees hotter than those in the suburbs. This, in turn, increased the number of local storms. Phoenix is also warmer than its outlying areas. In 1950, Phoenix was 6 degrees warmer than the nearby Casa Grande Monument. By 2007, however, the temperature difference increased to 14 degrees.

When trees grow throughout urban areas, both surface and air temperatures are reduced. Researchers have found that planting one tree to the west and one to the south of a home can significantly reduce energy consumption. In the Environmental Protection Agency's study, annual cooling costs were reduced by 8 to 18 percent while annual heating costs were reduced 2 to 8 percent.

Leaves also filter particles from the air, including dust, ozone, carbon monoxide and other air pollutants. Through the process of photosynthesis, trees remove carbon dioxide (a greenhouse gas) and release oxygen into our air. Trees store the carbon dioxide, called carbon sequestration, and — depending on the size of the tree — can hold between 35 to 800 pounds of carbon dioxide each year.

Trees aren't our saviors from smog, though. Photochemical smog is smog caused when sunlight and chemical compounds such as car exhaust combine. Trees contribute to this when they release organic gases.

Additionally, planting trees as a solution to global warming — a practice commonly linked to carbon offsets — may have a positive impact on global temperature control only when planted in the tropics, a thin geographical belt around the equator. Normally, trees help cool the planet by absorbing carbon dioxide as part of the photosynthesis process and by evaporating water into the air. In the tropics, water evaporates naturally from trees, increasing cloud cover and keeping temperatures cooler. Outside of the tropics, however, researchers are finding that forests trap heat because their dense, dark canopies absorb sunlight.

12

Making and Maintaining

Benefits of Trees

Most trees and shrubs in cities or communities are planted to provide beauty or shade. These are two excellent reasons for their use. Woody plants also serve many other purposes, and it often is helpful to consider these other functions when selecting a tree or shrub for the landscape. The benefits of trees can be grouped into social, communal, environmental, and economic categories.

Social Benefits

We like trees around us because they make life more pleasant. Most of us respond to the presence of trees beyond simply observing their beauty. We feel serene, peaceful, restful, and tranquil in a grove of trees. We are "at home" there. Hospital patients have been shown to recover from surgery more quickly when their hospital room offered a view of trees. The strong ties between people and trees are most evident in the resistance of community residents to removing trees to widen streets. Or we note the heroic efforts of individuals and organisations to save particularly large or historic trees in a community.

The stature, strength, and endurance of trees give them a cathedral-like quality. Because of their potential for long life, trees frequently are planted as living memorials. We often become personally attached to trees that we or those we love have planted.

Communal Benefits

Even though trees may be private property, their size often makes them part of the community as well. Because trees occupy considerable space, planning is required if both you and your neighbours are to benefit. With proper selection and maintenance, trees can enhance and function on one property without infringing on the rights and privileges of neighbours.

City trees often serve several architectural and engineering functions. They provide privacy, emphasize views, or screen out objectionable views. They reduce glare and reflection. They direct pedestrian traffic. They provide background to and soften, complement, or enhance architecture.

Environmental Benefits

Trees alter the environment in which we live by moderating climate, improving air quality, conserving water, and harboring wildlife. Climate control is obtained by moderating the effects of sun, wind, and rain. Radiant energy from the sun is absorbed or deflected by leaves on deciduous trees in the summer and is only filtered by branches of deciduous trees in winter. We are cooler when we stand in the shade of trees and are not exposed to direct sunlight. In winter, we value the sun's radiant energy. Therefore, we should plant only small or deciduous trees on the south side of homes.

Wind speed and direction can be affected by trees. The more compact the foliage on the tree or group of trees, the greater the influence of the windbreak. The downward fall of rain, sleet, and hail is initially absorbed or deflected by trees, which provides some protection for people, pets, and buildings. Trees intercept water, store some of it, and reduce storm runoff and the possibility of flooding.

Dew and frost are less common under trees because less radiant energy is released from the soil in those areas at night.

Temperature in the vicinity of trees is cooler than that away from trees. The larger the tree, the greater the cooling. By using trees in the cities, we are able to moderate the heat-island effect caused by pavement and buildings in commercial areas. Air quality can be improved through the use of trees, shrubs, and turf. Leaves filter the air we breathe by removing dust and other particulates. Rain then washes the pollutants to the ground. Leaves absorb carbon dioxide from the air to form carbohydrates that are used in the plant's structure and function. In this process, leaves also absorb other air pollutants—such as ozone, carbon monoxide, and sulfur dioxide—and give off oxygen.

By planting trees and shrubs, we return to a more natural, less artificial environment. Birds and other wildlife are attracted to the area. The natural cycles of plant growth, reproduction, and decomposition are again present, both above and below ground. Natural harmony is restored to the urban environment.

Economic Benefits

Individual trees and shrubs have value, but the variability of species, size, condition, and function makes determining their economic value difficult. The economic benefits of trees can be both direct and indirect. Direct economic benefits are usually associated with energy costs. Air-conditioning costs are lower in a tree-shaded home. Heating costs are reduced when a home has a windbreak. Trees increase in value from the time they are planted until they mature. Trees are a wise investment of funds because landscaped homes are more valuable than nonlandscaped homes. The savings in energy costs and the increase in property value directly benefit each home owner.

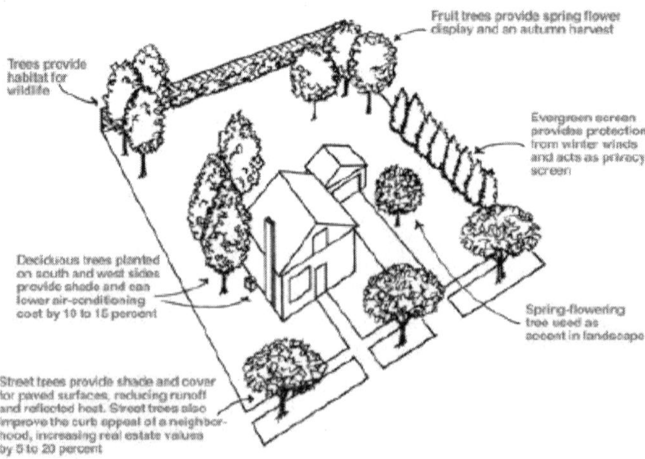

The indirect economic benefits of trees are even greater. These benefits are available to the community or region. Lowered electricity bills are paid by customers when power companies are able to use less water in their cooling towers, build fewer new facilities to meet peak demands, use reduced amounts of fossil fuel in their furnaces, and use fewer measures to control air pollution. Communities also can save money if fewer facilities must be built to control storm water in the region. To the individual, these savings are small, but to the community, reductions in these expenses are often in the thousands of dollars.

Trees Require an Investment

Trees provide numerous aesthetic and economic benefits but also incur some costs. You need to be aware that an investment is required for your trees to provide the benefits that you desire. The biggest cost of trees and shrubs occurs when they are purchased and planted. Initial care almost always includes some watering. Leaf, branch, and whole tree removal and disposal can be expensive.

To function well in the landscape, trees require maintenance. Much can be done by the informed home owner. Corrective pruning and mulching gives trees a good start. Shade trees, however, quickly grow to a size that may require the services of a professional arborist. Arborists have the knowledge and equipment needed to prune, spray, fertilize, and otherwise maintain a large tree. Your garden centre owner, university extension agent, community forester, or consulting arborist can answer questions about tree maintenance, suggest treatments, or recommend qualified arborists.

The PHC Alternative

Maintaining mature landscapes is a complicated undertaking. You may wish to consider a professional plant health care (PHC) maintenance program that is now available from many landscape care companies. The program is designed to maintain plant vigor and initially should include inspections to detect and treat any existing problems that could be damaging or fatal. Thereafter, regular inspections and preventive maintenance help ensure plant health and beauty.

Tree Values

Almost everyone knows that trees and other living plants are valuable. They beautify our surroundings, purify our air, act as sound barriers, manufacture precious oxygen, and help us save energy through their cooling shade in summer and their wind reduction in winter. Many people don't realise, however, that plants have a dollar value of their own that can be measured by competent plant appraisers.

If your trees or shrubs are damaged or destroyed, you may be able to recapture your loss through an insurance claim or as a deduction from your federal income tax.

Some Practical Advice

Here is some practical advice that may help you find out what your trees and plants are worth (a process known as valuation).

Planning for Highest Value

A professional in the tree, nursery, or landscape industry can help you plan, develop, install, and care for all of your trees and plants so that each of them will be worth more to you.

How Your Trees and Shrubs Are Valuated

Seek the advice of professionals in this industry who have developed a set of guidelines for the valuation. Such guidelines have been widely adopted in the field and are recognised by insurance companies, the courts, and, in some cases, the Internal Revenue Service (IRS).

What to Do If You Suffer Loss or Damage to Your Landscape Plants?

A casualty loss is defined by the IRS as "... a loss resulting from an identifiable event of sudden, unexpected, or unusual nature." This definition can include such events as vehicular accidents, storms, floods, lightning, vandalism, or even air and soil pollution.

If you suffer damage to trees or landscaping from any type of casualty, first consult your home owner's insurance policy to determine the amount and kind of coverage. Contact the insurance company to have an appraisal made by a competent tree and landscape professional who is experienced in plant appraisal. Have the appraisal made right after your loss or damage.

The tree and landscape appraiser accomplishes many things for you. The professional can see things you might miss, help correct damage, and prescribe remedies you may be able to do yourself. The appraiser will establish the amount of your loss in financial terms, including the cost of removing debris and making repairs as well as replacements. All of these steps are wise investments and well worth the cost you may incur for the inspection.

Four Factors in Professional Valuation of Trees and Other Plants

Size. Sometimes the size and age of a tree are such that it cannot be replaced. Trees that are too large to be replaced should be assessed by professionals who use a specialised appraisal formula.

Species or classification. Trees that are hardy, durable, highly adaptable, and free from objectionable characteristics are most valuable. They require less maintenance; they have sturdy, well-shaped branches, and pleasing foliage. Tree values vary according to your region, the "hardiness" zone, and even state and local conditions. If you are not familiar with these variables, be sure your advice comes from a competent source.

Condition. The professional will also consider the condition of the plant. Obviously, a healthy, well-maintained plant has a higher value. Roots, trunk, branches, and buds need to be inspected

Location. Functional considerations are important. A tree in your yard may be worth more than one growing in the woods. A tree standing alone often has a higher value than one in a group. A tree near your house or one that is a focal point in your landscape tends to have more value. The site, placement, and contribution of a tree to the overall landscape help determine the overall value of the plant attributable to location.

All of these factors can be measured in dollars and cents. They can determine the value of a tree, specimen shrubs, or evergreens, whether for insurance purposes, court testimony in lawsuits, or tax deductions.

Checklist

These steps should be taken before and after any casualty loss to your trees and landscape. Taking them can improve the value of your investment in nature's green, growing gifts and prevent financial loss should they be damaged or destroyed.

- Plan your landscaping for both beauty and functional value.
- Protect and preserve to maintain value.
- Take pictures of trees and other landscape plants now while they are healthy and vigorous. Pictures make "before and after" comparisons easier and expedite the processing of insurance claims or deductions for losses on federal tax forms.
- Check your insurance. In most cases, the amount of an allowable claim for any one tree or shrub is a maximum of $500.
- For insurance, legal, and income tax purposes, keep accurate records of your landscape and real estate appraisals on any losses.
- Consult your local Plant Health Care professional at every stage in the life cycle of your landscape (planning, planting, care) and to make sure you do not suffer needless financial loss when a casualty strikes.

Tree Selection

Tree selection is one of the most important investment decisions a home owner makes when landscaping a new home or replacing a tree lost to damage or disease. Considering that most trees have the potential to outlive the people who plant them, the impact of this decision is one that can influence a lifetime. Match the tree to the site, and both lives will benefit. The question most frequently asked of tree care professionals is "Which kind of tree do you think I should plant?" Before this question can be answered, a number of factors need to be considered. Think about the following questions:

- Why is the tree being planted? Do you want the tree to provide shade, fruit, or seasonal colour, or act as a windbreak or screen? Maybe more than one reason?
- What is the size and location of the planting site? Does the space lend itself to a large, medium, or small tree? Are there overhead or belowground wires or utilities in the vicinity? Do you need to consider clearance for sidewalks, patios, or driveways? Are there other trees in the area? Are there barriers to future root growth, such as building foundations?

- Which type of soil conditions exist? Is the soil deep, fertile, and well drained, or is it shallow, compacted, and infertile?
- Which type of maintenance are you willing to provide? Do you have time to water, fertilize, and prune the newly planted tree until it is established, or will you be relying on your garden or tree service for assistance?

Asking and answering these and other questions before selecting a tree will help you choose the "right tree for the right place."

Tree Function

Trees make our surroundings more pleasant. Properly placed and cared for, trees increase the value of our real estate. A large shade tree provides relief from summer's heat and, when properly placed, can reduce summer cooling costs. An ornamental tree provides beautiful flowers, leaves, bark, or fruit. Evergreens with dense, persistent leaves can be used to provide a windbreak or a screen for privacy. A tree that drops its leaves in the fall allows the sun to warm a house in the winter. A tree or shrub that produces fruit can provide food for the owner and/or attract birds and wildlife into your home landscape. Street trees decrease the glare from pavement, reduce runoff, filter out pollutants, and add oxygen to the air we breath. Street trees also improve the overall appearance and quality of life in a city or neighbourhood.

Form and Size

A basic principle of modern architecture is "form follows function." This is a good rule to remember when selecting a tree. Selecting the right form (shape) to complement the desired function (what you want the tree to do) can significantly reduce maintenance costs and increase the tree's value in the landscape. When making a selection about form, also consider mature tree size. Trees grow in a variety of sizes and shapes, as shown below. They can vary in height from several inches to several hundred feet. Select a form and size that will fit the planting space provided.

Depending on your site restrictions, you can choose from among hundreds of combinations of form and size. You may choose a small-spreading tree in a location with overhead utility lines. You may select a narrow, columnar form to provide a screen between two buildings. You may choose large, vase-shaped trees to create an arbor over a driveway or city street. You may even determine that the site just does not have enough space for a tree of any kind.

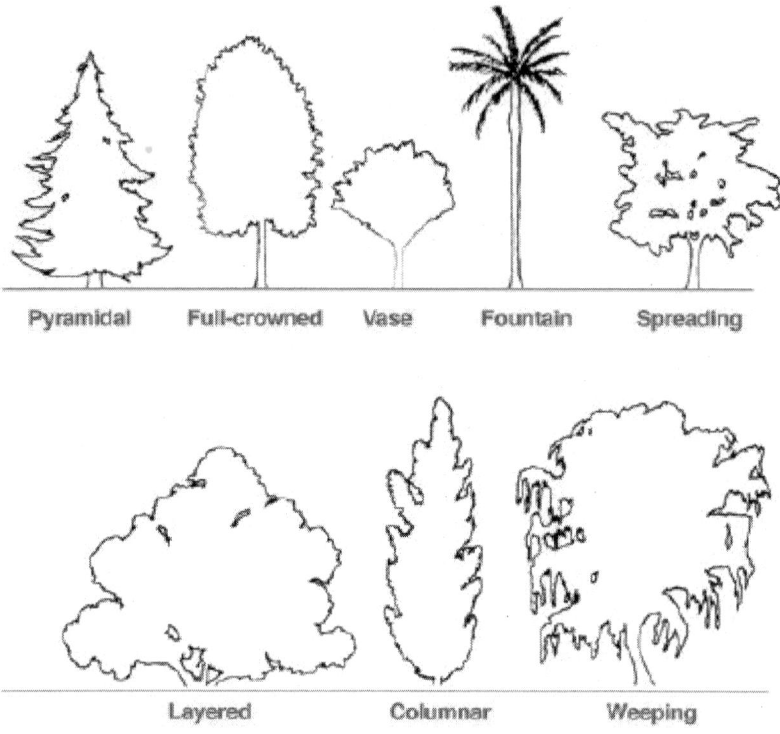

Site Conditions

Selecting a tree that will thrive in a given set of site conditions is the key to long-term tree survival. The following is a list of the major site conditions to consider before selecting a tree for planting:

- soil conditions
- exposure (sun and wind)
- human activity
- drainage
- space constraints
- hardiness zone

Soil Conditions

The amount and quality of soil present in your yard can limit planting success. In urban sites, the topsoil often has been disturbed and frequently is shallow, compacted, and subject to drought. Under these conditions, trees are continuously under stress. For species that are not able to handle these types of conditions, proper maintenance designed to reduce stress is necessary to ensure adequate growth and survival. Many arborists will, for a minor charge, take soil samples from your yard to test for fertility, salinity, and pH (alkalinity or acidity). The tests will be returned with recommendations on ways to improve poor soil conditions with fertilizers or soil amendments (sand, compost, or manure) and will also help your local nursery or garden centre recommend tree species that will do well in the soils found on your site.

Exposure

The amount of sunlight available will affect tree and shrub species selection for a particular location. Most woody plants require full sunlight for proper growth and flower bloom. Some do well in light shade, but few tree species perform well in dense shade. Exposure to wind is also a consideration. Wind can dry out soils, causing drought conditions and damage to branches and leaves during storms, and can actually uproot newly planted trees that have not had an opportunity to establish root systems. Special maintenance, such as staking or more frequent watering, may be needed to establish young trees on windy sites.

Human Activity

This aspect of tree selection is often overlooked. The reality of the situation is that the top five causes of tree death are the result of things people do: soil compaction, underwatering, overwatering, vandalism, and the number one cause—planting the wrong tree—account for more tree deaths than all insect and disease-related tree deaths combined.

Drainage

Tree roots require oxygen to develop and thrive. Poor drainage can remove the oxygen available to the roots from the soil and kill the tree. Before planting, dig some test holes 12 inches wide by 12 inches deep in the areas you are considering planting trees. Fill the holes with water and time how long it takes for the water to drain away. If it takes more than 6 hours, you may have a drainage problem. If so, ask your local garden centre for recommendations on how to correct the problem, or choose a different site.

Space Constraints

Many different factors can limit the planting space available to the tree: overhead or underground utilities, pavement, buildings, other trees, visibility. The list goes on and on. Make sure there is adequate room for the tree you select to grow to maturity, both above and below ground.

Hardiness

Hardiness is the plant's ability to survive in the extreme temperatures of the particular geographic region in which you are planting the tree. Plants can be cold hardy, heat tolerant, or both. Most plant reference books provide a map of hardiness zone ranges. Although tropical areas are generally Zone 11, higher elevations have cooler temperatures that may warrant adjustment to the hardiness zone classification. Check with your local garden centre for the hardiness information for your region. Before you make your final decision, make sure the plant you have selected is "hardy" in your area.

Pest Problems

Insect and disease organisms affect almost every tree and shrub species. Every plant has its particular pest problems, and the severity varies geographically. These pests may or may not be life threatening to the plant. You should select plants resistant to pest problems for your area. Your local ISA Certified Arborist, tree consultant, or extension agent can direct you to information relevant to problem species for your location.

Species Selection

Personal preferences play a major role in the selection process. Now that your homework is done, you are ready to select a species for the planting site you have chosen. Make sure you use the information you have gathered about your site conditions, and balance it with the aesthetic decisions you make related to your personal preferences.

The species must be suitable for the geographic region (hardy), tolerant to the moisture and drainage conditions of your soil, be resistant to pests in your area, and have the right form and size for the site and function you have envisioned.

Remember, the beautiful picture of a tree you looked at in a magazine or book was taken of a specimen that is growing vigorously because it was planted in the right place. If your site conditions tell you the species you selected will not do well under those conditions, do not be disappointed when the tree does not perform in the same way. If you are having difficulty answering any of these questions on your own, contact a local ISA Certified Arborist, tree care professional, garden centre, or extension agent for assistance. Their assistance will help you to plant the "right tree in the right place." It is better to get a professional involved early and help you make the right decision than to call him or her later to ask if you made the wrong decision.

Buying High-Quality Trees

When you buy a high-quality tree, plant it correctly, and treat it properly, you and your tree will benefit greatly in many ways for many years. When you buy a low-quality tree, you and your tree will have many costly problems even if you take great care in planting and maintenance.

What Determines Tree Quality?

A high-quality tree has

- enough sound roots to support healthy growth.
- a trunk free of mechanical wounds and wounds from incorrect pruning.
- a strong form with well-spaced, firmly attached branches.

A Low-Quality Tree has

- crushed or circling roots in a small root ball or small container.
- a trunk with wounds from mechanical impacts or incorrect pruning.
- a weak form in which multiple stems squeeze against each other or branches squeeze against the trunk.

Any of these problems alone or in combination with the others will greatly reduce the tree's chances for a long, attractive, healthy, and productive life.

When buying a tree, inspect it carefully to make certain it does not have problems with roots, injuries, or form. Remember the acronym RIF; it will help you remember roots, injuries, and f orm.

Here are some details on potential problems and some other considerations that you should be aware of when buying a tree.

Root Problems

Roots on trees for sale are available as one of three types:

- bare root: no soil; usually on small trees
- root balled: roots in soil held in place by burlap or some other fabric; the root ball may be in a wire basket
- container grown: roots and soil in a container

Bare-Root Stock

Bare roots should not be crushed or torn. The ends of the roots should be clean cut. If a few roots are crushed, re-cut them to remove the injured portions. Use sharp tools. Make straight cuts. Do not paint the ends. The cuts should be made immediately before planting and watering.

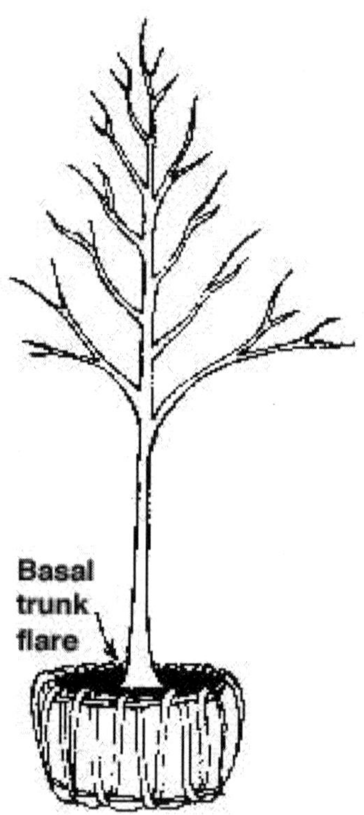

Root-Balled Stock

You should be able to see the basal trunk flare. The flare is the spreading trunk base that connects with the roots. Root balls should be flat on top. Roots in soil in round bags often have many major woody roots cut or torn during the bagging process. Avoid trees with many crushed or torn roots.

The diameter of the root ball should be at least 10 to 12 times the diameter of the trunk as measured 6 inches above the trunk flare.

After placing the root ball in the planting site, cut the ties and carefully pull away the burlap or other fabric. Examine any roots that protrude from the soil. If many roots are obviously crushed or torn, the tree may have severe growth problems. If only a few roots are injured, cut away only the injured portions. Use a sharp tool. Use care not to break the soil ball around the roots.

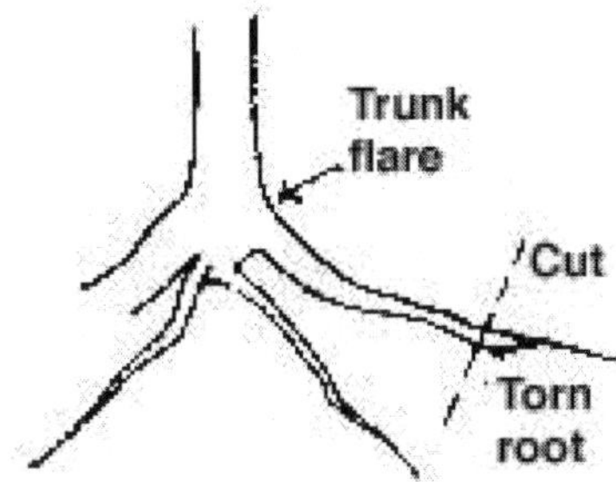

Cut the wire on wire baskets. Place the basket into the planting site. Cut away at least the top two wires without disturbing the root ball. Inspect exposed roots for injuries. If many roots are injured, the tree may have serious growth problems. If the trunk flare has been buried, gently expose it before planting the tree, taking care not to damage the bark.

Container-Grown Stock

Roots should not twist or circle in the container. Remove the root ball from the container. Inspect the exposed larger roots carefully to see whether they are twisting or turning in circles. Circling roots often girdle and kill other roots. If only a few roots are circling, cut them away with a sharp tool.

Trunk flare should be obvious. Be on alert for trees planted too deeply in containers or trees "buried" in fabric bags. As with root-balled stock, you should be able to see the basal trunk flare with container-grown plants. If the trunk flare has been buried, gently expose it before planting the tree, taking care not to damage the bark.

Injuries

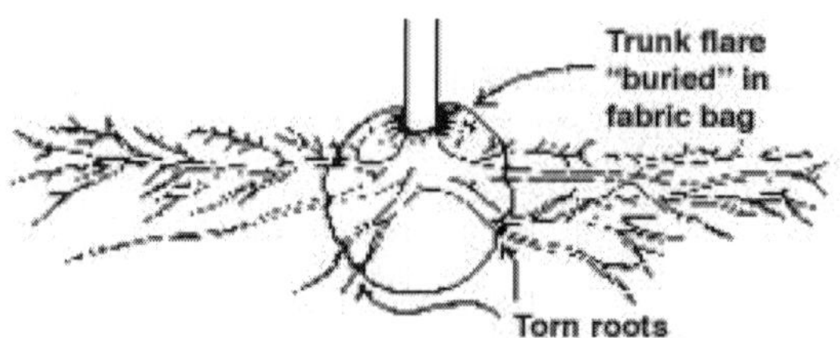

Small, Round Ball; Torn Roots

Beware of injuries beneath trunk wraps. Never buy a tree without thoroughly checking the trunk. If the tree is wrapped, remove the wrap and inspect the trunk for wounds, incorrect pruning cuts, and insect injuries. Wrap can be used to protect the trunk during transit but should be removed after planting.

Incorrect pruning cuts are major problems. Incorrect pruning cuts that remove or injure the swollen collar at the base of branches can start many serious tree problems, cankers, decay, and cracks.

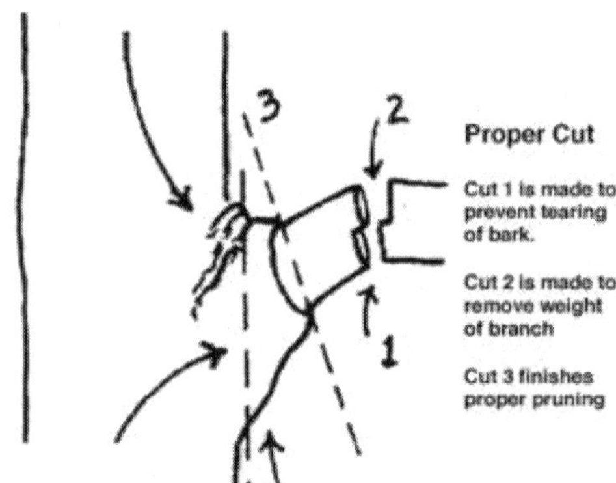

Incorrect pruning cuts that leave branch and leader stubs also start disease and defect problems. Do not leave stubs.

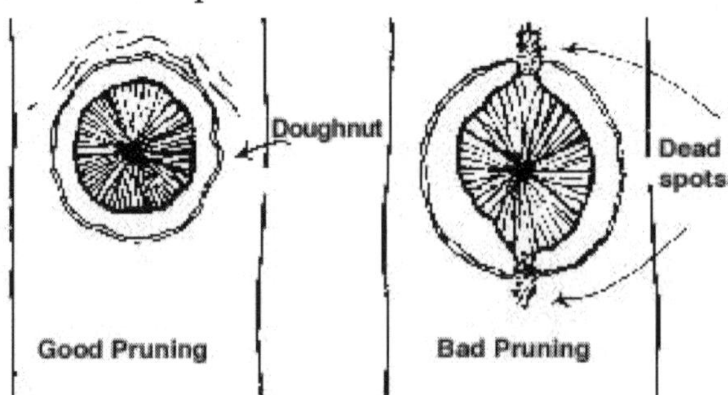

A correct pruning cut removes the branch just outside of the collar. A ring, or "doughnut," of sound tissues then grows around the cut. Do not make cuts flush to the trunk. The closing tissues may form only to the sides of the flush cuts. Trunk tissues above and below flush cut branches often die. When the heat of the sun or the cold of frost occurs, cracks or long, dead streaks may develop above and below the dead spots.

Form

Good, strong form, or architecture, starts with branches evenly spaced along the trunk. The branches should have firm, strong attachments with the trunk.

Squeezed branches signal problems. Weak branch unions occur where the branch and trunk squeeze together. As the squeezing increases during diameter growth, dead spots or cracks often begin to form below where the branch is attached to the trunk. Once this problem starts, the weak branch attachment could lead to branches cracking or breaking during mild to moderate storms.

When several branches are on the same position on the trunk, the likelihood of weak attachments and cracks increases greatly. As the branches grow larger and tighter together, the chances for splitting increase.

Avoid trees with two or more stems squeezing together. As stems squeeze together, cracks often form down the trunk. The cracks could start from squeezed multiple leader stems or where the two trunks come together.

If you desire a tree with multiple trunks, make certain that the trunks are well separated at the ground line.

Remember, trunks expand in diameter as they grow. Two trunks may be slightly separated when small, but as they grow in girth, the trunks will squeeze together.

Look for early signs of vertical trunk cracks. Examine branch unions carefully for small cracks below the unions. Cracks are major starting points for fractures of branches and trunks. The small cracks could be present for many years before a fracture happens. Always keep a close watch for vertical cracks below squeezed branches and squeezed trunks.

If your tree has only a few minor problems, corrective pruning may help. Start corrective pruning one year after planting. Space the pruning over several years. Remove broken or torn branches at the time of planting. After a year, start corrective pruning by removing the branches that died after planting.

Trees Have Dignity, Too

Most nurseries produce high-quality trees. When you start with a high-quality tree, you are giving that tree a chance to express its dignity for many years. Remember RIF.

Avoiding Tree & Utility Conflicts

Determining where to plant a tree is a decision that should not be taken lightly. Many factors should be considered prior to planting. When planning what type of tree to plant, remember to look up and look down to determine where the tree will be located in relation to overhead and underground utility lines.

Often, we take utility services for granted because they have become a part of our daily lives. For us to enjoy the convenience of reliable, uninterrupted service, distribution systems are required to bring utilities into our homes. These services arrive at our homes through overhead or underground lines.

Overhead lines can be electric, telephone, or cable television. Underground lines include those three plus water, sewer, and natural gas.

The location of these lines should have a direct impact on your tree and planting site selection. The ultimate mature height of a tree to be planted must be within the available overhead growing space. Just as important, the soil area must be large enough to accommodate the particular rooting habits and ultimate trunk diameter of the tree. Proper tree and site selection provide trouble-free beauty and pleasure for years to come.

Overhead Lines

Overhead utility lines are the easiest to see and probably the ones we take most for granted. Although these lines look harmless enough, they can be extremely dangerous. Planting tall-growing trees under and near these lines eventually requires your utility to prune them to maintain safe clearance from the wires. This pruning may result in the tree having an unnatural appearance. Periodic pruning can also lead to a shortened life span for the tree. Trees that must be pruned away from power lines are under greater stress and are more susceptible to insects and disease. Small, immature trees planted today can become problem trees in the future.

Tall-growing trees near overhead lines can cause service interruptions when trees contact wires. Children or adults climbing in these trees can be severely injured or even killed if they come in contact with the wires. Proper selection and placement of trees in and around overhead utilities can eliminate potential public safety hazards, reduce expenses for utilities and their rate payers, and improve the appearance of landscapes.

Underground Lines

Trees are much more than just what you see overhead. Many times, the root area is larger than the branch spread above ground. Much of the utility service provided today runs below ground. Tree roots and underground lines often coexist without problems. However, trees planted near underground lines could have their roots damaged if the lines need to be dug up for repairs.

The biggest danger to underground lines occurs during planting. Before you plant, make sure that you are aware of the location of any underground utilities. To be certain that you do not accidentally dig into any lines and risk serious injury or a costly service interruption, call your utility company or utility protection service first. Never assume that these utility lines are buried deeper than you plan to dig. In some cases, utility lines are very close to the surface.

Proper Places for Trees around Homes

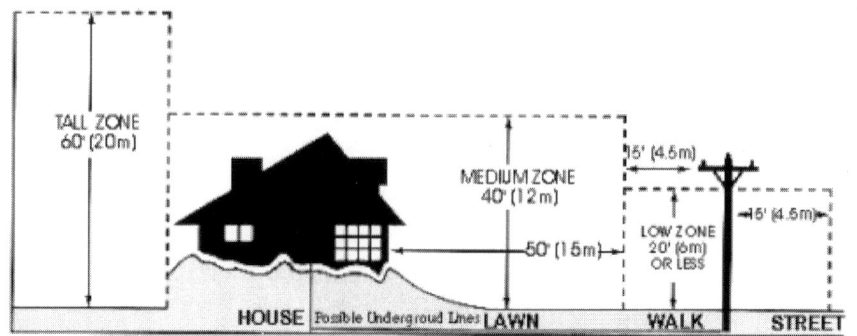

The illustration indicates approximately where trees should be planted in relation to utility lines. Your garden centre staff or tree care professional will gladly help you select the right tree.

Tall Zone

Trees that grow as tall as 60 feet (20 metres) can be used in the area marked as the tall zone; however, you should consider your neighbour's view or their existing plantings of flower beds and/or trees.

Plant large trees at least 35 feet (11 metres) away from the house for proper root development and to minimize damage to the house or building. These large-growing trees are also recommended for streets without overhead restrictions.

Street planting sites must also have wide planting areas or medians [greater than 8 feet (3 metres)] that allow for a large root system, trunk diameter, and root flare.

Large trees are also recommended for parks, meadows, or other open areas where their large size, both above and below ground, will not be restricted, cause damage, or become a liability.

Medium Zone

Trees that grow up to 40 feet (12 metres) tall can be used to decorate or frame your house or provide a parklike setting. Select your trees first, then plant shrubs to complement the trees. Medium-sized trees are also recommended for planting anywhere the available above and below ground growing space will allow them to reach a mature height of 30 to 40 feet (10 to 12 metres). Appropriate soil spaces are wide planting areas or medians [4 to 8 feet (1 to 3 metres) wide], large planting squares [8 feet (3 metres) square or greater], and other open areas of similar size or larger.

Low Zone

This zone extends 15 feet (4.5 metres) on either side of the wires. Trees with a mature height of less than 20 feet (6 metres) may be planted anywhere within this zone, including street tree plantings under utility lines. Such trees are also recommended when the growing space is limited. These trees are appropriate as well for narrow planting areas [less than 4 feet (1 metre) wide]; planting squares or circles surrounded by concrete; large, raised planting containers; or other locations where underground space for roots will not support tall- or medium-zone trees.

Some Further Suggestions

Plant evergreen trees to serve as windbreaks on the west or north side of the house, approximately 50 feet (15 metres) or more from the house.

Plant deciduous trees (those that drop their leaves in the fall) on the south and/or west side of the house to cool in the summer and allow sun to enter the house in the winter.

New Tree Planting

Think of the tree you just purchased as a lifetime investment. How well your tree, and investment, grows depends on the type of tree and location you select for planting, the care you provide when the tree is planted, and follow-up care the tree receives after planting.

Planting the Tree

The ideal time to plant trees and shrubs is during the dormant season and in the fall after leaf drop or early spring before budbreak. Weather conditions are cool and allow plants to establish roots in the new location before spring rains and summer heat stimulate new top growth. However, trees properly cared for in the nursery or garden centre, and given the appropriate care during transport to prevent damage, can be planted throughout the growing season. In tropical and subtropical climates where trees grow year round, any time is a good time to plant a tree, provided that sufficient water is available. In either situation, proper handling during planting is essential to ensure a healthy future for new trees and shrubs. Before you begin planting your tree, be sure you have had all underground utilities located prior to digging.

If the tree you are planting is balled or bare root, it is important to understand that its root system has been reduced by 90 to 95 percent of its original size during transplanting. As a result of the trauma caused by the digging process, trees commonly exhibit what is known as transplant shock. Containerized trees may also experience transplant shock, particularly if they have circling roots that must be cut. Transplant shock is indicated by slow growth and reduced vigor following transplanting. Proper site preparation before and during planting coupled with good follow-up care reduces the amount of time the plant experiences transplant shock and allows the tree to quickly establish in its new location. Carefully follow nine simple steps, and you can significantly reduce the stress placed on the plant at the time of planting.

1. Dig a shallow, broad planting hole. Make the hole wide, as much as three times the diameter of the root ball but only as deep as the root ball. It is important to make the hole wide because the roots on the newly establishing tree must push through surrounding soil in order to establish. On most planting sites in new developments, the existing soils have been compacted and are unsuitable for healthy root growth. Breaking up the soil in a large area around the tree provides the newly emerging roots room to expand into loose soil to hasten establishment.
2. Identify the trunk flare. The trunk flare is where the roots spread at the base of the tree. This point should be partially visible after the tree has been planted. If the trunk flare is not partially visible, you may have to remove some soil from the top of the root ball. Find it so you can determine how deep the hole needs to be for proper planting.
3. Remove tree container for containerized trees. Carefully cutting down the sides of the container may make this easier. Inspect the root ball for circling roots and cut or remove them. Expose the trunk flare, if necessary.
4. Place the tree at the proper height. Before placing the tree in the hole, check to see that the hole has been dug to the proper depth and no more. The majority of the roots on the newly planted tree will develop in the top 12 inches of soil. If the tree is planted too deeply, new roots will have difficulty developing because of a lack of oxygen. It is better to plant the tree a little high, 2 to 3 inches above the base of the trunk flare, than to plant it at or below the original growing level. This planting level will allow for some settling. To avoid damage when setting the tree in the hole, always lift the tree by the root ball and never by the trunk.

5. Straighten the tree in the hole. Before you begin backfilling, have someone view the tree from several directions to confirm that the tree is straight. Once you begin backfilling, it is difficult to reposition the tree.
6. Fill the hole gently but firmly. Fill the hole about one-third full and gently but firmly pack the soil around the base of the root ball. Then, if the root ball is wrapped, cut and remove any fabric, plastic, string, and wire from around the trunk and root ball to facilitate growth. Be careful not to damage the trunk or roots in the process.

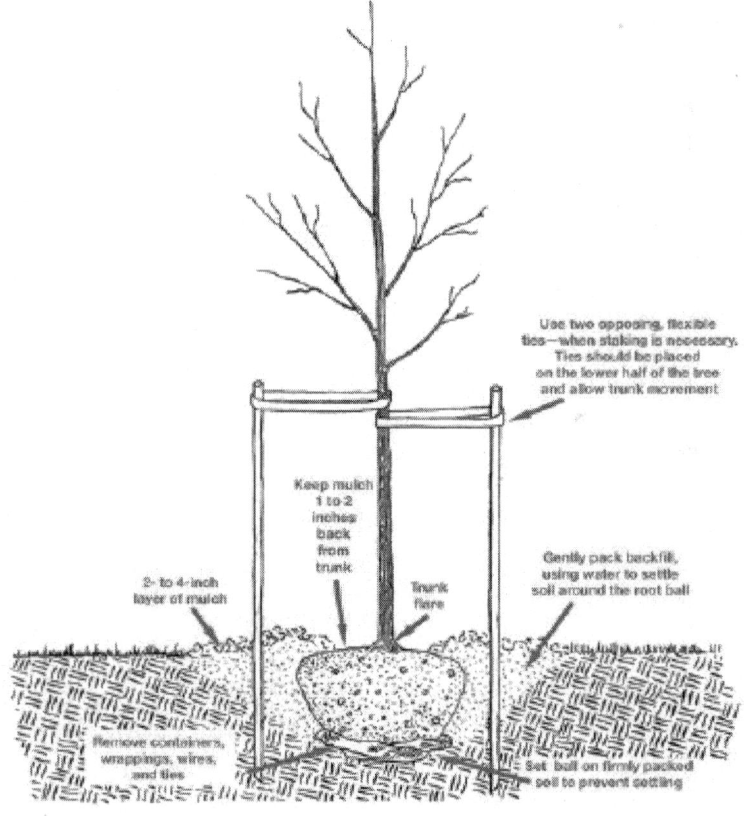

Fill the remainder of the hole, taking care to firmly pack soil to eliminate air pockets that may cause roots to dry out. To avoid this problem, add the soil a few inches at a time and settle with water. Continue this process until the hole is filled and the tree is firmly planted. It is not recommended to apply fertilizer at the time of planting.

7. Stake the tree, if necessary. If the tree is grown and dug properly at the nursery, staking for support will not be necessary in most home landscape situations. Studies have shown that trees establish more quickly and develop stronger trunk and root systems if they are not staked at the time of planting. However, protective staking may be required on sites where lawn mower damage, vandalism, or windy conditions are concerns. If staking is necessary for support, there are three methods to choose among: staking, guying, and ball stabilizing. One of the most common methods is staking. With this method, two stakes used in conjunction with a wide, flexible tie material on the lower half of the tree will hold the tree upright, provide flexibility, and minimize injury to the trunk. Remove support staking and ties after the first year of growth.
8. Mulch the base of the tree. Mulch is simply organic matter applied to the area at the base of the tree. It acts as a blanket to hold moisture, it moderates soil temperature extremes, and it reduces competition from grass and weeds. Some good choices are leaf litter, pine straw, shredded bark, peat moss, or composted wood chips. A 2- to 4-inch layer is ideal. More than 4 inches may cause a problem with oxygen and moisture levels. When placing mulch, be sure that the actual trunk of the tree is not covered. Doing so may cause decay of the living bark at the base of the tree. A mulch-free area, 1 to 2 inches wide at the base of the tree, is sufficient to avoid moist bark conditions and prevent decay.
9. Provide follow-up care. Keep the soil moist but not soaked; overwatering causes leaves to turn yellow or fall off. Water trees at least once a week, barring rain, and more frequently during hot weather. When the soil is dry below the surface of the mulch, it is time to water. Continue until mid-fall, tapering off for lower temperatures that require less-frequent watering.

Other follow-up care may include minor pruning of branches damaged during the planting process. Prune sparingly immediately after planting and wait to begin necessary corrective pruning until after a full season of growth in the new location.

After you have completed these nine simple steps, further routine care and favourable weather conditions will ensure that your new tree or shrub will grow and thrive. A valuable asset to any landscape, trees provide a long-lasting source of beauty and enjoyment for people of all ages. When questions arise about the care of your tree, be sure to consult your local ISA Certified Arborist or a tree care or garden centre professional for assistance.

Mature Tree Care

Think of tree care as an investment. A healthy tree increases in value with age—paying big dividends, increasing property values, beautifying our surroundings, purifying our air, and saving energy by providing cooling shade from summer's heat and protection from winter's wind.

Providing a preventive care program for your landscape plants is like putting money in the bank. Regular maintenance, designed to promote plant health and vigor, ensures their value will continue to grow. Preventing a problem is much less costly and time-consuming than curing one once it has developed. An effective maintenance program, including regular inspections and the necessary follow-up care of mulching, fertilizing, and pruning, can detect problems and correct them before they become damaging or fatal. Considering that many tree species can live as long as 200 to 300 years, including these practices when caring for your home landscape is an investment that will offer enjoyment and value for generations.

Tree Inspection

Tree inspection is an evaluation tool to call attention to any change in the tree's health before the problem becomes too serious. By providing regular inspections of mature trees at least once a year, you can prevent or reduce the severity of future disease, insect, and environmental problems. During tree inspection, be sure to examine four characteristics of tree vigor: new leaves or buds, leaf size, twig growth, and absence of crown dieback (gradual death of the upper part of the tree).

A reduction in the extension of shoots (new growing parts), such as buds or new leaves, is a fairly reliable cue that the tree's health has recently changed. To evaluate this factor, compare the growth of the shoots over the past three years. Determine whether there is a reduction in the tree's typical growth pattern.

Further signs of poor tree health are trunk decay, crown dieback, or both. These symptoms often indicate problems that began several years before. Loose bark or deformed growths, such as trunk conks (mushrooms), are common signs of stem decay.

Any abnormalities found during these inspections, including insect activity and spotted, deformed, discoloured, or dead leaves and twigs, should be noted and watched closely. If you are uncertain as to what should be done, report your findings to your local ISA Certified Arborist or other tree care professional for advice on possible treatment.

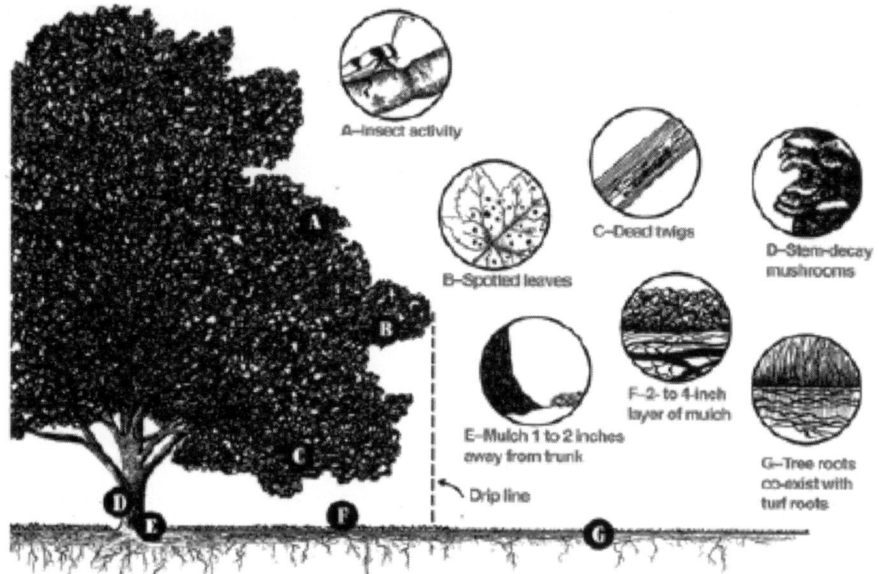

Mulching

Mulching can reduce environmental stress by providing trees with a stable root environment that is cooler and contains more moisture than the surrounding soil. Mulch can also prevent mechanical damage by keeping machines such as lawn mowers and string trimmers away from the tree's base. Further, mulch reduces competition from surrounding weeds and turf.

To be most effective in all of these functions, mulch should be placed 2 to 4 inches deep and cover the entire root system, which may be as far as 2 or 3 times the diameter of the branch spread of the tree. If the area and activities happening around the tree do not permit the entire area to be mulched, it is recommended that you mulch as much of the area under the drip line of the tree as possible (refer to diagram). When placing mulch, care should be taken not to cover the actual trunk of the tree. This mulch-free area, 1 to 2 inches wide at the base, is sufficient to avoid moist bark conditions and prevent trunk decay.

An organic mulch layer 2 to 4 inches deep of loosely packed shredded leaves, pine straw, peat moss, or composted wood chips is adequate. Plastic should not be used because it interferes with the exchange of gases between soil and air, which inhibits root growth. Thicker mulch layers, 5 to 6 inches deep or greater, may also inhibit gas exchange.

Fertilization

Fertilization is another important aspect of mature tree care. Trees require certain nutrients (essential elements) to function and grow. Urban landscape trees can be growing in soils that do not contain sufficient available nutrients for satisfactory growth and development. In these situations, it may be necessary to fertilize to improve plant vigor.

Fertilizing a tree can improve growth; however, if fertilizer is not applied wisely, it may not benefit the tree at all and may even adversely affect the tree. Mature trees making satisfactory growth may not require fertilization. When considering supplemental fertilizer, it is important to know which nutrients are needed and when and how they should be applied.

Soil conditions, especially pH and organic matter content, vary greatly, making the proper selection and use of fertilizer a somewhat complex process. When dealing with a mature tree that provides considerable benefit and value to your landscape, it is worth the time and investment to have the soil tested for nutrient content. Any arborist can arrange to have your soil tested at a soil testing laboratory and can give advice on application rates, timing, and the best blend of fertilizer for each of your trees and other landscape plants.

Mature trees have expansive root systems that extend from 2 to 3 times the size of the leaf canopy. A major portion of actively growing roots is located outside the tree's drip line. It is important to understand this fact when applying fertilizer to your trees as well as your turf. Many lawn fertilizers contain weed and feed formulations that may be harmful to your trees. When you apply a broadleaf herbicide to your turf, remember that tree roots coexist with turf roots. The same herbicide that kills broadleaf weeds in your lawn is picked up by tree roots and can harm or kill your broadleaf trees if applied incorrectly. Understanding the actual size and extent of a tree's root system before you fertilize is necessary to determine how much, what type, and where to best apply fertilizer.

Pruning

Pruning is the most common tree maintenance procedure next to watering. Pruning is often desirable or necessary to remove dead, diseased, or insect-infested branches and to improve tree structure, enhance vigor, or maintain safety. Because each cut has the potential to change the growth of (or cause damage to) a tree, no branch should be removed without a reason.

Removing foliage from a tree has two distinct effects on its growth. Removing leaves reduces photosynthesis and may reduce overall growth. That is why pruning should always be performed sparingly. Overpruning is extremely harmful because without enough leaves, a tree cannot gather and process enough sunlight to survive.

However, after pruning, the growth that does occur takes place on fewer shoots, so they tend to grow longer than they would without pruning. Understanding how the tree responds to pruning should assist you when selecting branches for removal.

Pruning mature trees may require special equipment, training, and experience. If the pruning work requires climbing, the use of a chain or hand saw, or the removal of large limbs, then using personal safety equipment, such as protective eyewear and hearing protection, is a must. Arborists can provide a variety of services to assist in performing the job safely and reducing risk of personal injury and damage to your property.

They also are able to determine which type of pruning is necessary to maintain or improve the health, appearance, and safety of your trees.

Removal

Although tree removal is a last resort, there are circumstances when it is necessary. An arborist can help decide whether or not a tree should be removed. Professionally trained arborists have the skills and equipment to safely and efficiently remove trees. Removal is recommended when a tree

- is dead, dying, or considered irreparably hazardous
- is causing an obstruction or is crowding and causing harm to other trees and the situation is impossible to correct through pruning
- is to be replaced by a more suitable specimen
- should be removed to allow for construction

With proper maintenance, trees are attractive and can add considerable value to your property. Poorly maintained trees, on the other hand, can be a significant liability. Pruning or removing trees, especially large trees, can be dangerous work. It should be performed only by those trained and equipped to work safely in trees.

Plant Health Care

The most common reason a tree owner calls an arborist is concern that something is wrong with a tree. It may be that some of the leaves are discoloured, a branch has died, or perhaps the entire tree has been dropping leaves. Sometimes the cause of concern is a minor problem that is easily explained and corrected. Other times the problem is more complex—with several underlying causes and a remedy that requires treatments extending over several years. Unfortunately, there are instances in which the problem has gone undetected for so long that the tree cannot be helped, and the only option is removal. If an arborist had been called earlier, perhaps the tree could have been saved.

The Solution: Plant Health Care

Situations such as these led arborists to create Plant Health Care (PHC) programs. The objective of PHC is to maintain or improve the landscape's appearance, vitality and—in the case of trees—safety, using the most cost-effective and environmentally sensitive practices and treatments available. Plant Health Care involves monitoring, using preventive treatments, and adopting a strong commitment to working closely with you, the tree owner.

Why Plant Health Care, Not Tree Health Care?

While trees are dominant ornamental features in your home landscape, they share this area with turfgrasses, shrubs, and bedding plants. And all these plants have one resource in common: the soil. The roots of trees, shrubs, turfgrass, and bedding plants intermingle and compete for water and nutrients. In fact, the roots of a single mature tree may extend 60 feet or more out into your lawn or flower beds. Every treatment applied to the lawn (fertilizer and herbicide, for example) can impact the appearance and vitality of a tree. Conversely, treatments applied to a tree, such as pruning and fertilizing, can influence the appearance and vitality of the underlying turfgrass. *The care of each plant in a landscape can affect the health of every plant in that landscape.*

Why Contact an Arborist for Plant Health Care?

Trees and shrubs represent a considerable long-term investment in your landscape. With proper care, these plants can provide beautiful surroundings, cooling shade, and many other benefits for decades. Arborists have the experience and training to detect many potential tree and shrub problems before they become life threatening or hazardous. In addition, arborists can make tree and shrub recommendations, such as species selection and placement, to keep many problems from occurring in the first place. Arborists can also consult with other landscape services you may use, lawn care for example, to ensure that the treatments are coordinated and will not be harmful to your trees and shrubs. *Remember, the potential size and longevity of trees and shrubs warrants their special attention in your landscape. Bedding plants can be replaced in a few short weeks and a lawn in a single growing season, but it can take a lifetime or more to replace a mature tree.*

What does a Tree and Shrub PHC Program Cover?

Every home landscape is unique, so there is no standard PHC program. Plant Health Care programs do have features in common, however. First, PHC involves monitoring tree and shrub health. This step allows problems to be detected and managed before they become serious. The monitoring may be as simple as annual visits to check on a few special trees in your landscape, or it may involve more frequent quarterly or monthly inspections of all your trees and shrubs. The monitoring frequency and complexity of your PHC program depend on the size and diversity of your landscape as well as your particular landscape goals.

Second, if problems or potential problems are detected or anticipated during a monitoring visit, your arborist will develop solutions. The solution could be a simple change in your lawn irrigation schedule—many trees are kept too moist—or more detailed suggestions, such as pruning or spot applications of pesticides.

Finally, PHC involves you, the client. Your arborist will give you information about your trees and shrubs. This information ensures that decisions are made that address your concerns and are appropriate to your landscape budget and goals. Information may be provided through a variety of means. Obviously, discussions and answering questions are important means of conveying information, but many PHC programs include written recommendations after each monitoring visit. *Plant Health Care is a program tailored to the needs of the client and his or her trees and shrubs.*

How will My Trees and Shrubs Benefit from PHC?

Because ornamental trees and shrubs can quickly succumb to problems, routine monitoring and timely treatments can protect your landscape investment and reduce expenses. A monitoring visit to your landscape might reveal

- a hidden infestation of tent caterpillars that may soon defoliate the ornamental crabapples in your front yard
- a weakly attached branch that may fail and damage the house
- improperly pruned shrubs that are not flowering as abundantly as they should.

Your Plant Health Care specialist can recommend treatments and changes in maintenance practices that can eliminate these problems while maximizing the safety and aesthetic quality of your landscape.

What will a PHC Program Cost?

Because each program is individually designed to fit the needs of a particular landscape, no standard price can be given without a site visit and assessment. You may have an interest in developing a plan for a few key trees in your landscape, or you may wish to have the entire landscape placed on a program. PHC programs can also be structured in different ways.

For example, some programs charge a fee for monitoring and bill each treatment separately. Other programs have an annual fee that covers all monitoring visits for the season as well as many potential treatments.

These more comprehensive programs provide the peace of mind in knowing that treatments for most potential problems are already covered by the program without additional charges. Individualized programs and flexibility are at the heart of PHC. You will find that your arborist can design a Plant Health Care program that fits your goals and budget.

Proper Mulching Techniques

Mulches are materials placed over the soil surface to maintain moisture and improve soil conditions. Mulching is one of the most beneficial things a home owner can do for the health of a tree. Mulch can reduce water loss from the soil, minimize weed competition, and improve soil structure. Properly applied, mulch can give landscapes a handsome, well-groomed appearance. Mulch must be applied properly; if it is too deep or if the wrong material is used, it can actually cause significant harm to trees and other landscape plants.

Benefits of Proper Mulching

- Helps maintain soil moisture. Evaporation is reduced, and the need for watering can be minimized.
- Helps control weeds. A 2- to 4-inch layer of mulch will reduce the germination and growth of weeds.
- Mulch serves as nature's insulating blanket. Mulch keeps soils warmer in the winter and cooler in the summer.
- Many types of mulch can improve soil aeration, structure (aggregation of soil particles), and drainage over time.
- Some mulches can improve soil fertility.
- A layer of mulch can inhibit certain plant diseases.
- Mulching around trees helps facilitate maintenance and can reduce the likelihood of damage from "weed whackers" or the dreaded "lawn mower blight."
- Mulch can give planting beds a uniform, well-cared-for look.

Trees growing in a natural forest environment have their roots anchored in a rich, well-aerated soil full of essential nutrients. The soil is blanketed by leaves and organic materials that replenish nutrients and provide an optimal environment for root growth and mineral uptake. Urban landscapes, however, are typically a much harsher environment with poor soils, little organic matter, and large fluctuations in temperature and moisture. Applying a 2- to 4-inch layer of organic mulch can mimic a more natural environment and improve plant health.

The root system of a tree is not a mirror image of the top. The roots of most trees can extend out a significant distance from the tree trunk. Although the guideline for many maintenance practices is the drip line—the outermost extension of the canopy—the roots can grow many times that distance. In addition, most of the fine, absorbing roots are located within inches of the soil surface. These roots, which are essential for taking up water and minerals, require oxygen to survive. A thin layer of mulch, applied as broadly as practical, can improve the soil structure, oxygen levels, temperature, and moisture availability where these roots grow.

Types of Mulch

Mulches are available commercially in many forms. The two major types of mulch are inorganic and organic. Inorganic mulches include various types of stone, lava rock, pulverized rubber, geotextile fabrics, and other materials. Inorganic mulches do not decompose and do not need to be replenished often. On the other hand, they do not improve soil structure, add organic materials, or provide nutrients. For these reasons, most horticulturists and arborists prefer organic mulches.

Organic mulches include wood chips, pine needles, hardwood and softwood bark, cocoa hulls, leaves, compost mixes, and a variety of other products usually derived from plants. Organic mulches decompose in the landscape at different rates depending on the material and climate. Those that decompose faster must be replenished more often. Because the decomposition process improves soil quality and fertility, many arborists and other landscape professionals consider that characteristic a positive one, despite the added maintenance.

Not Too Much!

As beneficial as mulch is, too much can be harmful. The generally recommended mulching depth is 2 to 4 inches. Unfortunately, many landscapes are falling victim to a plague of overmulching. A new term, "mulch volcanoes," has emerged to describe mulch that has been piled up around the base of trees. Most organic mulches must be replenished, but the rate of decomposition varies. Some mulches, such as cypress mulch, remain intact for many years. Top dressing with new mulch annually (often for the sake of refreshing the colour) creates a buildup to depths that can be unhealthy. Deep mulch can be effective in suppressing weeds and reducing maintenance, but it often causes additional problems.

Problems Associated with Improper Mulching

- Deep mulch can lead to excess moisture in the root zone, which can stress the plant and cause root rot.

- Piling mulch against the trunk or stems of plants can stress stem tissues and may lead to insect and disease problems.
- Some mulches, especially those containing cut grass, can affect soil pH. Continued use of certain mulches over long periods can lead to micronutrient deficiencies or toxicities.
- Mulch piled high against the trunks of young trees may create habitats for rodents that chew the bark and can girdle the trees.
- Thick blankets of fine mulch can become matted and may prevent the penetration of water and air. In addition, a thick layer of fine mulch can become like potting soil and may support weed growth.
- Anaerobic "sour" mulch may give off pungent odors, and the alcohols and organic acids that build up may be toxic to young plants.

Proper Mulching

It is clear that the choice of mulch and the method of application can be important to the health of landscape plants. The following are some guidelines to use when applying mulch.

- Inspect plants and soil in the area to be mulched. Determine whether drainage is adequate. Determine whether there are plants that may be affected by the choice of mulch. Most commonly available mulches work well in most landscapes. Some plants may benefit from the use of a slightly acidifying mulch such as pine bark.
- If mulch is already present, check the depth. Do not add mulch if there is a sufficient layer in place. Rake the old mulch to break up any matted layers and to refresh the appearance. Some landscape maintenance companies spray mulch with a water-soluble, vegetable-based dye to improve the appearance.
- If mulch is piled against the stems or tree trunks, pull it back several inches so that the base of the trunk and the root crown are exposed.
- Organic mulches usually are preferred to inorganic materials due to their soil-enhancing properties. If organic mulch is used, it should be well aerated and, preferably, composted. Avoid sour-smelling mulch.
- Composted wood chips can make good mulch, especially when they contain a blend of leaves, bark, and wood. Fresh wood chips also may be used around established trees and shrubs.

Avoid using noncomposted wood chips that have been piled deeply without exposure to oxygen.

- For well-drained sites, apply a 2- to 4-inch layer of mulch. If there are drainage problems, a thinner layer should be used. Avoid placing mulch against the tree trunks. Place mulch out to the tree's drip line or beyond.

Remember: If the tree had a say in the matter, its entire root system (which usually extends well beyond the drip line) would be mulched.

Pruning Mature Trees

Pruning is the most common tree maintenance procedure. Although forest trees grow quite well with only nature's pruning, landscape trees require a higher level of care to maintain their safety and aesthetics. Pruning should be done with an understanding of how the tree responds to each cut. Improper pruning can cause damage that will last for the life of the tree, or worse, shorten the tree's life.

Reasons for Pruning

Because each cut has the potential to change the growth of the tree, no branch should be removed without a reason. Common reasons for pruning are to remove dead branches, to remove crowded or rubbing limbs, and to eliminate hazards. Trees may also be pruned to increase light and air penetration to the inside of the tree's crown or to the landscape below. In most cases, mature trees are pruned as a corrective or preventive measure.

Routine thinning does not necessarily improve the health of a tree. Trees produce a dense crown of leaves to manufacture the sugar used as energy for growth and development. Removal of foliage through pruning can reduce growth and stored energy reserves. Heavy pruning can be a significant health stress for the tree.

Yet if people and trees are to coexist in an urban or suburban environment, then we sometimes have to modify the trees. City environments do not mimic natural forest conditions. Safety is a major concern. Also, we want trees to complement other landscape plantings and lawns. Proper pruning, with an understanding of tree biology, can maintain good tree health and structure while enhancing the aesthetic and economic values of our landscapes.

When to Prune

Most routine pruning to remove weak, diseased, or dead limbs can be accomplished at any time during the year with little effect on the tree. As a rule, growth is maximized and wound closure is fastest if pruning takes place before the spring growth flush. Some trees, such as maples and birches, tend to "bleed" if pruned early in the spring. It may be unsightly, but it is of little consequence to the tree.

A few tree diseases, such as oak wilt, can be spread when pruning wounds allow spores access into the tree. Susceptible trees should not be pruned during active transmission periods. Heavy pruning just after the spring growth flush should be avoided. At that time, trees have just expended a great deal of energy to produce foliage and early shoot growth. Removal of a large percentage of foliage at that time can stress the tree.

Making Proper Pruning Cuts

Pruning cuts should be made just outside the branch collar. The branch collar contains trunk or parent branch tissue and should not be damaged or removed. If the trunk collar has grown out on a dead limb to be removed, make the cut just beyond the collar. Do not cut the collar.

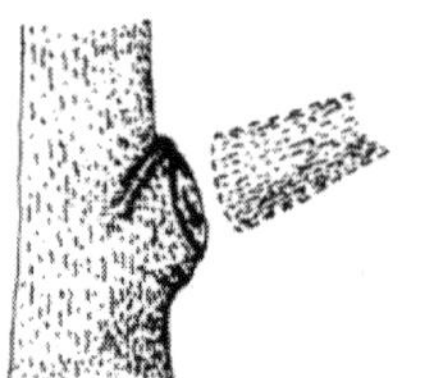

Pruning cuts should be made just outside the branch collar.

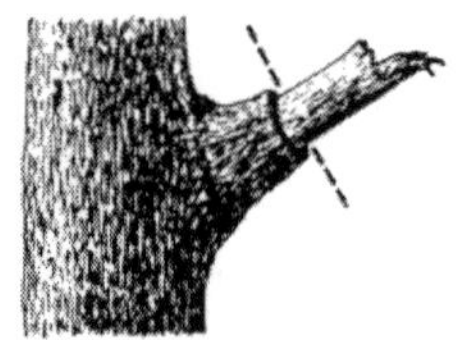

On a dead branch that has a collar of live wood, the final cut should be made just beyond the outer edge of the collar

If a large limb is to be removed, its weight should first be reduced. This is done by making an undercut about 12 to 18 inches from the limb's point of attachment. Make a second cut from the top, directly above or a few inches farther out on the limb. Doing so removes the limb, leaving the 12- to 18-inch stub. Remove the stub by cutting back to the branch collar. This technique reduces the possibility of tearing the bark.

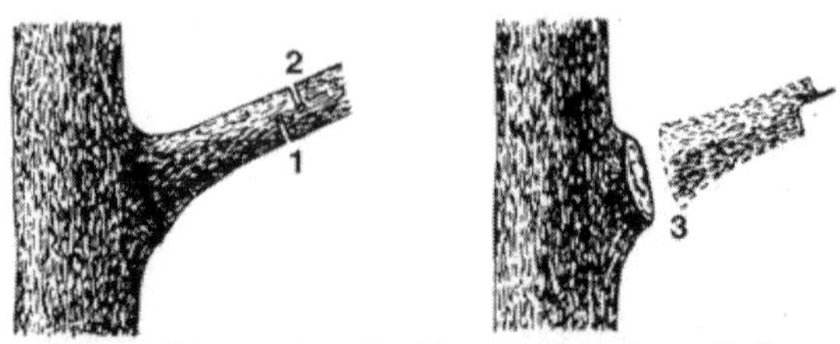

Use the three-cut method to remove a large limb.

Pruning Techniques

Specific types of pruning may be necessary to maintain a mature tree in a healthy, safe, and attractive condition.

Cleaning is the removal of dead, dying, diseased, crowded, weakly attached, and low-vigor branches from the crown of a tree.

Thinning is the selective removal of branches to increase light penetration and air movement through the crown. Thinning opens the foliage of a tree, reduces weight on heavy limbs, and helps retain the tree's natural shape.

Raising removes the lower branches from a tree in order to provide clearance for buildings, vehicles, pedestrians, and vistas.

Reduction reduces the size of a tree, often for clearance for utility lines. Reducing the height or spread of a tree is best accomplished by pruning back the leaders and branch terminals to lateral branches that are large enough to assume the terminal roles (at least one-third the diameter of the cut stem). Compared to topping, reduction helps maintain the form and structural integrity of the tree.

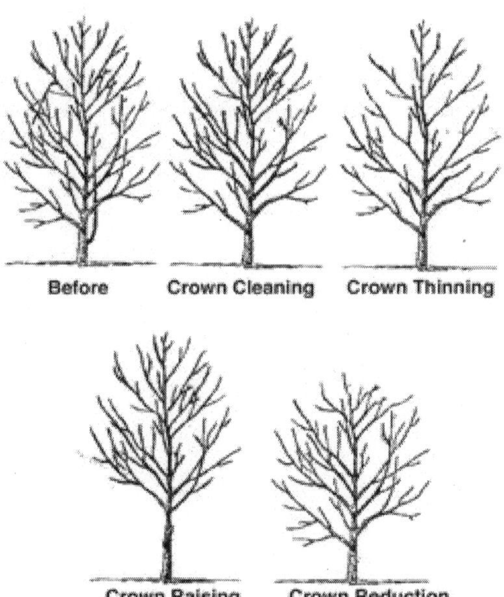

How Much should be Pruned?

The amount of live tissue that should be removed depends on the tree size, species, and age, as well as the pruning objectives. Younger trees tolerate the removal of a higher percentage of living tissue better than mature trees do. An important principle to remember is that a tree can recover from several small pruning wounds faster than from one large wound.

A common mistake is to remove too much inner foliage and small branches. It is important to maintain an even distribution of foliage along large limbs and in the lower portion of the crown. Overthinning reduces the tree's sugar production capacity and can create tip-heavy limbs that are prone to failure.

Mature trees should require little routine pruning. A widely accepted rule of thumb is never to remove more than one-quarter of a tree's leaf-bearing crown. In a mature tree, pruning even that much could have negative effects. Removing even a single, large-diameter limb can create a wound that the tree may not be able to close. The older and larger a tree becomes, the less energy it has in reserve to close wounds and defend against decay or insect attack. The pruning of large mature trees is usually limited to removal of dead or potentially hazardous limbs.

Wound Dressings

Wound dressings were once thought to accelerate wound closure, protect against insects and diseases, and reduce decay. However, research has shown that dressings do not reduce decay or speed closure and rarely prevent insect or disease infestations. Most experts recommend that wound dressings not be used. If a dressing must be used for cosmetic purposes, then only a thin coating of a nontoxic material should be applied.

Hiring an Arborist

Pruning large trees can be dangerous. If pruning involves working above the ground or using power equipment, it is best to hire a professional arborist. An arborist can determine the type of pruning necessary to improve the health, appearance, and safety of your trees. A professional arborist can provide the services of a trained crew, with all of the required safety equipment and liability insurance.

There are a variety of things to look for when selecting an arborist:

- membership in professional organisations such as the International Society of Arboriculture (ISA), the Tree Care Industry Association (TCIA), or the American Society of Consulting Arborists (ASCA)
- certification through ISA's Certified Arborist program
- proof of insurance
- list of references (don't hesitate to check)

Avoid using the services of any tree company that

- advertises topping as a service provided; knowledgeable arborists know that topping is harmful to trees and is not an accepted practice
- uses tree climbing spikes to climb trees that are being pruned; climbing spikes can damage trees, and their use should be limited to trees that are being removed

Helping Trees Recover from Stress

Trees are the dominant component in the natural and managed landscape in New England. For the most part, trees in forests thrive and, typically, live more than one hundred years. On the other hand, trees planted in cities and towns, and along roadways, often survive no more than a few decades, if that long.

These trees are subjected to a variety of factors that reduce vigor and may eventually lead to decline and death. Almost all of these factors are the result of human activity. It is the purpose of this fact sheet to introduce the reader to the stresses imposed on trees growing in communities and around homes, describe the common symptoms of stress, and suggest ways to prevent or reduce the impact of these stresses on trees.

Causes of Stress: An Overview

Most people believe that insects and diseases are the primary cause for decline and death of trees in the landscape. In fact, it is human activity which causes most of the problems that trees experience.

Even many pest and disease problems can be related directly or indirectly to the prior stresses imposed upon trees by human activity.

These activities may include improper planting and pruning of trees, as well as poor care. Applying too much or too little water, especially after planting, may injure roots and cause stress. Over application of fertilizer can also lead to root injury or rapid but weak growth that makes the tree prone to certain insect pests and plant diseases. Mechanical injuries from lawn mowers and weed trimmers can damage tree bark and create the opportunity for invasion by certain disease organisms. Excessive foot or vehicular traffic around trees can destroy the structure of soil leading to compaction of that soil. Compaction reduces the level of oxygen in soil, slows the penetration of water, and hinders the development of roots.

Other consequences of human activity, especially air pollution and road salt applications, are a little more difficult to assess and to prevent. For people living in areas prone to frequent air pollution events, the best remedy, as far as tree health is concerned, is to select species that are somewhat tolerant of common air pollutants. Similarly, different species of trees have different tolerances to salts. If exposure to road salt is a problem, for example, salt tolerant species such as ginkgo may be planted.

Diagnosing the source of stress on a tree can sometimes be difficult because tree decline is rarely caused by one factor alone. It is more likely that a combination of factors are working together to cause the decline of a tree over a period of many years. The growing environment of a stressed tree should be carefully evaluated to determine as many of the causal factors as possible. Check for proper planting depth and for girdling roots. The past history of the tree should also be reviewed. Previous applications of chemicals, recent periods of drought, extremely cold or open winters, flooding and similar events can be important in evaluating sources of stress.

Nurseries and garden centres sell a great variety of trees, each with its own specific environmental requirements. Landscape features such as soil acidity, sun or shade, level of soil moisture and drainage can influence how well a tree will grow and prosper. It is important to match these site features with the environmental needs of a tree before making a purchase.

Symptoms of Stress

The symptoms of stress often develop slowly, more slowly than insect or disease symptoms. Some things to look for include: the appearance of abnormally small leaves, pale green colouration of leaves, unusually slow growth, premature development of fall leaf colouration, early leaf drop, dieback of twigs and branches, wilting of leaves and tender new growth, peeling bark, and presence of fungi protruding from stems and branches. Repeated occurrence of these symptoms over a period of years is a good indicator that a tree is being subjected to some chronic stress influences.

Some sources of stress, e.g. a prolonged drought, may reduce the vigor of a tree but, if provided with a good growing environment, the tree may overcome this stress and resume normal growth. However, it should be noted that it may take a tree several years to overcome a single stress factor. Also, the effects of drought on tree health may not become apparent until 2 or 3 years after the drought event.

Unfortunately, once symptoms of decline or stress become apparent, it is often too late to stop or reverse the damage to a tree. Therefore, prevention is the best approach to eliminating stress to trees. Prevention begins with proper selection and planting, and continues with regular inspection and care of these trees.

Proper Planting Techniques

Perhaps nothing is more important in preventing stress than the proper planting of trees. Research has shown that a tree planted at the correct depth, in a hole of sufficient size to accommodate the tree's expanding root system, has a much greater chance of survival than an improperly planted one. Here are some rules to follow when planting trees:

- Locate trees away from high foot traffic areas that may cause compaction of soil and away from the street if road salts are routinely applied during the winter. Avoid planting trees where their roots may be confined, such as close to a street, sidewalk or building. Also, avoid planting trees beneath utility wires.
- Evaluate the planting site with respect to drainage, exposure to wind, amount of sunlight, type of soil (sandy or clay), space for root development, and soil pH. Also, note the hardiness zone in which you live so that you are sure to select a tree that will survive the prevailing winter temperatures. Hardiness zones in Massachusetts range from USDA zone 7a on Cape Cod to USDA zone 4 in the Berkshires.
- Have soil tested prior to planting. Phosphorous, potassium and limestone are best incorporated into the soil before a tree is planted. When applied to the soil surface after planting,

limestone, phosphorous, and potassium (to some extent), move very slowly into the soil. A soil test will determine the amount, if any, of these materials that needs to be incorporated. It is not necessary to apply any other fertilizer at planting time. However, if fertilizer containing nitrogen is applied, the nitrogen must be in a slow-release or water insoluble form. Organic fertilizers are a good choice. Follow the directions on the product label. At planting time it is better to apply too little nitrogen than too much. Fertilizers should be mixed thoroughly with the backfill material.

- Select a tree whose growth requirements most closely match the environment of the site where the tree will be planted. Remember, choose the right plant for the right place.
- Dig the planting hole to a depth equal to the height of the root ball, as measured from the trunk flare to bottom of the ball. To determine the true height of a root ball that is wrapped in burlap, remove the twine and burlap at the top of the root ball. Then carefully remove soil away from the tree trunk until you find the point where the trunk begins to flare. This is the natural point where the tree trunk ends and the root system begins. Planting a tree with the flare below ground level can lead to stress and subsequent death of the tree.
- Dig the hole at least two to three times as wide as the width of the root ball. If the site soil is severely compacted, the hole should be four to five times the root ball width.
- Set the tree into the hole so that the top of the root ball, where the trunk flare is visible, is at the same level as the surrounding grade or 1 to 2 inches higher than grade. Leave the bottom of the hole undisturbed.
- After setting the tree in the planting hole, remove all the twine and as much of the burlap as possible without allowing the rootball to break apart. If non-degradable materials have been used in wrapping the rootball (such as plastic burlap), remove all of the material before backfilling. It is also advisable to remove the top 8 to 16 inches of a wire basket if one has been used around the rootball. Removal of the wire should only be done after the rootball is securely in place in the planting hole.
- Backfill the hole with the original soil. Do not amend the backfill with peat moss or other organic materials. If the soil is of very poor quality, that is, very sandy or of heavy clay, it may be amended prior to digging the hole by working organic

matter into an area that is at least 5 times larger than the diameter of the tree's rootball. Organic matter amendment should not exceed 20-35% of the backfill by volume or 5% by weight.

- Firm the backfill by lightly tamping the soil. *Do not pack the soil!* You may also firm the soil by stopping periodically and watering the backfill until the soil is settled.
- Water the soil thoroughly after completing the planting. Then apply a 2 to 4 inch layer of organic mulch, such as bark chips, over the planting area. *Do not place any of the mulch against the trunk of the tree!*
- Water the soil around newly planted trees once each week through the first growing season. Soil should be soaked to a depth of at least 6 inches at each watering. A slow soaking, extended over several hours is more effective than a large application of water all at once. Speciality "soaker" hoses are effective in applying water to newly planted trees.

Caring for Trees

Once trees are planted, some routine care should be given to them to insure their health and vigor. This includes pruning, watering, fertilizing, mulching and checking for presence of any insect or disease problems.

Pruning

Pruning is an important element in the care of trees. The primary purposes of pruning are to remove dead, damaged or diseased branches; to selectively eliminate crowded branches or stems; to correct structural problems with tree limbs and stems; and to control the shape of a tree.

Pruning should be initiated while trees are still young and small. It is much easier and safer to remove poorly growing stems and branches on a small trees than on older, larger ones. While proper pruning does require some knowledge and skill, much pruning work can be done by the homeowner.

Do corrective pruning to:

- prevent branches from rubbing.
- remove dead, diseased or damaged branches.
- remove double leaders.
- remove narrow, V-shaped crotches.
- remove branches growing back toward the centre of the tree.

Do not remove or "top" the central leader (stem) unless it is damaged.

Cut branches at the branch collar, not flush against the trunk.

Use the "three cut" technique for large branches.

If pruning requires the use of a chain saw, removal of very large branches, or climbing into the tree, it is best that this work be done by a certified arborist (check local Yellow Pages under "Tree Service"). Pruning under such situations can be very dangerous and should be left to the professional.

Watering

Once a tree is established, it seldom needs watering. The exceptions are for trees growing in sandy soils, and those exposed to a prolonged drought. When watering, it is important to apply enough water to penetrate the soil to a depth of at least 10 to 14 inches. This is where most of the water absorbing roots of a tree are located. A test hole can be dug to determine how long one needs to water, or how much water needs to be applied, to moisten the soil to that depth. In general, such applications should be made about once every two weeks.

Fertilizing

Trees growing in a forest do quite well by relying on nutrients made available from rocks in the soil, and from the decay of organic matter in the soil. In home landscapes, trees may not get all the nutrients they need by way of these processes. Therefore, it is sometimes necessary to apply fertilizer around trees to provide the essential elements needed for growth. It should not be assumed, however, that trees need annual applications of fertilizer for their survival. Too much fertilizer can actually lead to stress on a tree rather than relieve it. Trees that have been in place for at least two or three years have a well developed root system capable of absorbing adequate amounts of mineral nutrients from the soil. Fertilizer application to trees should only be made if tree leaves are showing symptoms of nutrient deficiency, such as abnormally pale leaf colour or abnormally small leaves. Other symptoms of nutrient deficiency include shorter than normal annual twig growth and tip die-back of branches. Be aware that these same symptoms may be the result of stress factors other than a lack of nutrients. Always evaluate the tree and the area around the tree for other possible explanations for the tree's decline.

If fertilizer is to be applied, use one with an N-P-K ratio of 3-1-2 or 4-1-2. Fertilizer with an analysis such as 24-8-16, 12-4-4, 18-

6-12 or similar formulation would be a good choice. Products designed for application to lawns are satisfactory but avoid those containing weed killers. Fertilizers containing a high proportion of the nitrogen in a slow-release or water insoluble form should be used.

Care should be taken when applying fertilizer or any garden chemical that correct amounts are used. Apply fertilizer at a rate that is equivalent to no more than 1 pound of actual nitrogen per 1000 square feet of area. This rate may be adjusted upward if the plant response is less than satisfactory. Fertilizer should be applied in spring or in early fall. The easiest way to apply fertilizer is to spread it on the ground around the tree, beginning about 6 feet away from the trunk and extending to a distance several feet beyond the drip line of the tree. Where there is danger of burning the lawn around a tree, fertilizer application may be split into 2 or 3 applications, each about 4 weeks apart. Water the area thoroughly after applying fertilizer.

Just as important as applying fertilizer is application of limestone. In Massachusetts, most soils are acidic, with the exception of parts of the Berkshires and planting areas near concrete surfaces. Limestone is used to reduce soil acidity. The amount to apply depends upon soil pH, a measure of soil acidity. Have soil tested for pH every 4 or 5 years. For the most part, trees prefer a pH between 6 and 7 for good growth.

Mulching

Mulching involves the placement of wood chips, bark nuggets, pine needles, or other organic material over the soil surface surrounding a tree. The purpose of the mulch is to keep soil cool during the hottest months, reduce moisture loss from the soil, and eliminate weeds that compete with trees for water and nutrients. The use of mulch can also help prevent injury to trees by keeping lawn mowers and weed trimmers away from the trunks of trees. Mulches should be applied to a depth of 2 to 4 inches and spread over as large a portion of the root zone of a tree as is possible.

The mulch should not be in contact with the trunk of the tree.

Since organic mulches decompose over time, it will be necessary to replenish the mulch every few years. However, the depth of the undecomposed organic mulch should not exceed 4 inches.

Insect/Disease Control

The key to insect and disease control in trees is to maintain a proper growing environment for trees and to encourage their vigor.

Vigor should not be interpreted as rapid growth since very fast growth, especially that induced by over fertilization, can make a tree more prone to pest problems. Good cultural practices such as those previously mentioned are crucial for vigorous growth.

Most trees are often able to withstand a certain amount of insect infestation. It is when pest populations build to very high levels that problems occur. Unfortunately, pests are not usually noticed until they are in abundance. Therefore, a program of regular and methodical inspection of trees should be implemented, especially during the spring months when many insect infestations occur. Tree inspection should include examination of bark, stems, and leaves for any signs of pests or abnormal appearance of these plant structures.

There are many options for controlling insect pests on trees, including many measures that do not rely on the application of chemical pesticides. When dealing with small trees, pest control materials may be applied by the homeowner. A local garden supply dealer may be consulted regarding the appropriate materials to use. With large trees, specialised equipment will be needed to make applications of control materials. In such cases, it may be practical to hire an arborist who is licensed and certified to apply pesticides.

Tree diseases develop slowly and are usually caused by fungi or bacteria. Symptoms of disease infections typically appear as leaf spots, blights, yellowing of leaves, or wilting. Proper recognition of tree disease symptoms requires familiarity with the "normal" appearance and growth of trees. Trees should also be examined regularly for signs of peeling bark, dieback of twigs and branches, and fungi growing out from the trunk or at the base of the trunk. Some diseases result in little lasting damage to trees while others become progressively worse and can eventually kill trees. The damage caused by diseases can be compounded if the tree is being affected by other stress factors such as drought. If a wilt disease or dieback of branches is occurring, the problem may be internal or below ground. In these situations, it is best to consult with a professional.

When leaf spot and blight symptoms are visible, it is usually too late to apply chemical controls for that season. In addition, trees with good vitality generally have the capacity to limit and compensate for these kinds of disease infections. When diseases occur, effective management strategies include removing fallen leaves, pruning diseased and damaged branches, and providing good care.

Bibliography

Aboukhaled A.: *Optical Properties of Leaves in Relation to their Energy-balance, Photosynthesis, and Water use Efficiency*, University of California, CA, 1966.

Ausubel, F.M. : *Current Protocols in Molecular Biology,* New York: John Wiley and Sons, 1989.

Bailey, L.H. : *Manual of Cultivated Plants*, The Macmillan Company, New York, 1949.

Banki, L.: *Bioassay of Pesticides in the Laboratory,* Akademiai Kiado, Budapest, 1978.

Banninger C.: *Laboratory and FLI Airborne Imaging Spectrometer Measurements of a Spruce Forest and their Relationship to Variations in Foliar Pigment Content*, Espoo, Finland, 1989.

Barnes, N.: *Biology*, New York, Worth Publishers, 1989.

Barnum, Susan R.: *Biotechnology: An Introduction*, Belmont, Thomson/Brooks/Cole, 2005.

Barrington, E. J. W.: *Biochemistry of Primitive Deuterostomians*, London, Academic Press, 1974.

Brandwein, P.F. : *Sourcebook for the Biological Sciences,* San Diego: Harcourt Brace Jovanovich, 1986.

Chaudhary, Vikas: *Entomology and Pest Management*, Navyug, Delhi, 2008.

Collymore L.: *Fruit Production in Barbados*, Port of Spain, Trinidad and Tobago, 1996.

Coste R.: *Coffee: the Plant and the Product*, London, MacMillan, 1992.

Daphne C. Elliott: *Biochemistry and Molecular Biology*, Oxford University Press, Delhi, 2005.

David Sadava: *Plants, Genes and Crop Biotechnology*, Sudbury MA, Jones and Barlett Publishers, 2003.

E. Ramann: *The Evolution and Classification of Soils*, Asiatic Pub, Calcutta, 2006.

Elliott, B N : *Biochemistry and Molecular Biology*, Oxford University Press, Delhi, 2005.

Featherly H. I.: *Taxonomic Terminology of the Higher Plants,* USA, Iowa State College Press, 1954.

Ganguly, Smriti : *Biochemistry of Biomolecules,* Pearl Books, Delhi, 2007.

Jeffers P.: *Evaluation of Four Onion Varieties in Montserrat,* Plymouth, CARDI, 1992.

Jones, R. M.: *Plant Resources of South-East Asia,* Wageningen, Pudoc Scientific Publishers, 1992.

Madan Lal Bagdi: *Physiology, Biochemistry and Biotechnology,* Manglam Pub, Delhi, 2007.

Madulid Domingo A.: *A Pictorial Cyclopedia of Philippine Ornamental Plants,* Philippines, Makati Metro Manila, 1995.

Mahindru, S. N.: *Food Safety and Pesticides,* APH, Delhi, 2009.

Muneesh Kainth: *Chordate Embryology,* Dominant, Delhi, 2003.

Nobel, P. S.: *Physicochemical and Environmental Plant Physiology,* Academic Press, San Diego, 1999.

Pemberton, R. W.: *Predictable Risk to Native Plants in Weed Biological Control,* Oecologia, 2000.

Qystein V. Sjaastad: *Physiology of Domestic Animals,* International Book Distributing Co., Delhi, 2005.

Ragone D.: *Breadfruit: Artocarpus Altilis (Parkinson) Fosberg,* Rome, International Plant Genetic Resources Institute, 1997.

Rifkin, Jeremy: *The Biotech Century,* New York, Penguin Putnam, 1998

Swarnim, K. : *A Textbook of Biochemistry and Microbiology,* Surendra Pub, Delhi, 2010.

Tawde, A. B.: *Propagation and Rootstocks of Mango,* New Delhi, Malhotra, 1993.

Urton, Gary: *The Social Life of Numbers,* Austin, University of Texas Press, 1997.

Vanangamudi, K.: *Principles and Methods of Plant Breeding,* International Book, Delhi, 2005.

Whealy K.: *The Garden Seed Inventory,* Decorah, Seed Saver Publications, 1988.

White, G.F.: *Natural Hazards: Local, National, Global,* Oxford University Press, New York, 1974.

Index

A

Air, 32, 84, 92, 96, 136, 138, 140, 274.
Allelopathy, 63.
Antifreeze Protein, 248.
Antitranspirant, 103.
Arabidopsis Thaliana, 248.
Associated Tree Damage, 267.
Atmosphere Continuum, 103, 134.
Autumn Leaf Colour, 58.

B

Basic Types, 22.
Benefits of Trees, 272.
Bulb-Scales, 55.
Butterfly, 239.

C

Carbon Fixation, 170, 202.
Cataphyll, 51, 52, 53.
Cell Walls, 62.
Chloroplast DNA, 154.
Cleaning Surfaces, 234, 239.
Climate Change, 96, 230, 262.

D

Deciduous, 3, 62, 65, 67.
Dehydrin, 254.
Density of Stomata, 87.
Desert Plants, 230, 231.
Diffusing Gases, 83.
DNA Replication, 155.

E

Emergence, 44.
Environmental Benefits, 273.

F

Ferns, 23, 24, 25, 26, 27, 28, 29, 30, 31, 32.
Frond, 32, 33, 34.
Functional Principle, 234.

G

Gas Exchange, 83, 84, 89, 95.

H

Hot Weather, 232.
Hydrophobic, 242.

I

Interactions, 19.
Interfacing, 140.

K

Keeping Cool, 228.

L

Leaf Loss, 8.
Leaking Water, 92.
Leaves, 1, 2, 4, 7, 8, 10, 15, 17, 21, 27, 69, 71, 72, 73, 77, 78, 80, 81, 85, 109, 141, 241, 243, 271, 272, 273, 276.

Lotus Effect, 234, 236, 238.
Lotus Leaves, 242.
Lycopodiophyta, 28, 40, 42, 43, 44.

M

Mature Tree Care, 295.
Mechanism Driving, 133.
Mechanisms of Action, 252.
Medium Zone, 290.

N

Nature Cleans, 237.

O

Omniphobic Surfaces, 242.
Organisms Confused, 32.
Oxygen Evolution, 200, 202.

P

Photodissociation, 195, 198.
Photorespiration, 171, 219, 220, 222, 224, 226, 227.
Photosynthesis, 7, 96, 140, 141, 143, 168, 195, 203, 208, 209, 210, 211, 212, 215.
Photosynthetic, 95, 141, 142, 189, 192, 200, 208, 215.
Pigments, 60, 69, 71, 164.

R

Raising Water, 121.
Regulation, 100, 175, 215.
Rice Leaves, 239.
Role of Endodermis, 121.
Roots, 27, 84, 88, 126, 127, 128, 138, 277, 283, 284, 285.

S

Seasonal Leaf Loss, 8.
Seeking Illumination, 58.
Soil Plant, 103, 134.
Starting the Story, 1.
Staying Stiff, 256.
Staying Unfrozen, 245.
Stoma, 85, 86, 87, 88, 92, 95, 96, 97, 98, 109, 111, 114, 116, 117, 118, 119, 120, 126.
Succulent Plant, 49.
Surviving a Storm, 267.

T

Thylakoid, 162, 163, 177, 178, 179, 180, 181, 182, 183, 185.
Transpiration Stream, 138.
Trees, 67, 69, 72, 105, 125, 267, 268, 269, 270, 271, 272, 273, 274, 275, 276, 277, 279, 283, 288, 289, 290, 291, 297, 300, 301, 302, 305, 309, 313, 314, 316.

V

Various Biofuel Crops, 211.
Vernalization, 246, 247, 248.

W

Water, 76, 79, 90, 92, 98, 99, 101, 104, 105, 106, 111, 112, 113, 115, 118, 121, 122, 123, 125, 128, 130, 133, 135, 136, 138, 168, 171, 183, 190, 191, 200, 201, 208, 211, 224, 228, 236, 237, 242, 243, 253, 259, 268, 286, 294, 310, 313, 314, 315, 316.
Woody Plants, 67.

❑❑❑